Security and Privacy Issues in IoT Devices and Sensor Networks

Security and Privacy Issues in IoT Devices and Sensor Networks

Volume Twelve

Series Editors

Nilanjan Dey
Amira S. Ashour
Simon James Fong

Edited by

Sudhir Kumar Sharma

Professor in the Department of Computer Science, Institute of Information Technology & Management, GGSIPU, New Delhi, India

Bharat Bhushan

Research Scholar of Computer Science and Engineering (CSE), Birla Institute of Technology, Mesra, Ranchi, Jharkhand, India

Narayan C. Debnath

Founding Dean, School of Computing and Information Technology, Eastern International University, Thu Dau Mot City, Vietnam

ELSEVIER

ACADEMIC PRESS

An imprint of Elsevier

Academic Press is an imprint of Elsevier
125 London Wall, London EC2Y 5AS, United Kingdom
525 B Street, Suite 1650, San Diego, CA 92101, United States
50 Hampshire Street, 5th Floor, Cambridge, MA 02139, United States
The Boulevard, Langford Lane, Kidlington, Oxford OX5 1GB, United Kingdom

Notices
Knowledge and best practice in this field are constantly changing. As new research and experience broaden our understanding, changes in research methods, professional practices, or medical treatment may become necessary.

Practitioners and researchers must always rely on their own experience and knowledge in evaluating and using any information, methods, compounds, or experiments described herein. In using such information or methods they should be mindful of their own safety and the safety of others, including parties for whom they have a professional responsibility.

To the fullest extent of the law, neither the Publisher nor the authors, contributors, or editors, assume any liability for any injury and/or damage to persons or property as a matter of products liability, negligence or otherwise, or from any use or operation of any methods, products, instructions, or ideas contained in the material herein.

Library of Congress Cataloging-in-Publication Data
A catalog record for this book is available from the Library of Congress

British Library Cataloguing-in-Publication Data
A catalogue record for this book is available from the British Library

ISBN 978-0-12-821255-4

For information on all Academic Press publications
visit our website at https://www.elsevier.com/books-and-journals

Publisher: Mara Conner
Acquisitions Editor: Fiona Geraghty
Editorial Project Manager: John Leonard
Production Project Manager: Prasanna Kalyanaraman
Cover Designer: Matthew Limbert

Typeset by SPi Global, India

Working together
to grow libraries in
developing countries

www.elsevier.com • www.bookaid.org

Contents

CHAPTER 8 Lightweight cryptographic algorithms for resource-constrained IoT devices and sensor networks 153

Pulkit Singh, Bibhudendra Acharya, and
Rahul Kumar Chaurasiya

Contributors

Bibhudendra Acharya
Department of Electronics and Communication Engineering, National Institute of Technology, Raipur, Chhattisgarh, India

A. Ananthi
School of Computing, Sathyabama Institute of Science and Technology, Chennai, India

B. Banuselvasaraswathy
Department of ECE, Sri Krishna College of Technology, Coimbatore, India

Gaurav Bathla
Computer Science & Engineering, Chandigarh University, Mohali, Punjab, India

Bharat Bhushan
Birla Institute of Technology, Ranchi, India

Boncho Bonev
Technical University of Sofia, Sofia, Bulgaria

Rahul Kumar Chaurasiya
Department of Electronics and Communication Engineering, Malaviya National Institute of Technology, Jaipur, Rajasthan, India

Narayan C. Debnath
School of Computing and Information Technology, Eastern International University, Thu Dau Mot City, Vietnam

D. Deepa
School of Computing, Sathyabama Institute of Science and Technology, Chennai, India

Mahdi Dhaini
Department of Computer Science and Mathematics, Lebanese American University, Beirut, Lebanon

Dinh-Thuan Do
Wireless Communications Research Group, Faculty of Electrical and Electronics Engineering, Ton Duc Thang University, Ho Chi Minh City, Vietnam

Amin Fakhereldine
Department of Computer Science and Mathematics, Lebanese American University, Beirut, Lebanon

Sukriti Goyal
HMR Institute of Technology & Management, Delhi, India

Ramzi A. Haraty
Department of Computer Science and Mathematics, Lebanese American University, Beirut, Lebanon

Lawrence Henesey
Department of CSE, Blekinge Institute of Technology, Karlskrona, Sweden

Mohammad Jaber
Department of Computer Science and Mathematics, Lebanese American University, Beirut, Lebanon

Igor Kabashkin
Transport and Telecommunication Institute, Riga, Latvia

Ila Kaushik
Krishna Institute of Engineering & Technology, Ghaziabad, Uttar Pradesh, India

Chanchal Kumar
Department of Computer Science, Jamia Millia Islamia, New Delhi, India

Rohit Kumar
Computer Science, Govt. P.G. College, Naraingarh, Haryana, India

Vivek Kumar
Noida Institute of Engineering and Technology, Greater Noida, India

Anh-Tu Le
Faculty of Electronics Technology, Industrial University of Ho Chi Minh City, Ho Chi Minh City, Vietnam

Priyanka Makkar
Amity University, Haryana, India

Alok Pradhan
Macquarie University, Sydney, NSW, Australia

Shiv Prakash
Department of Chemical Engineering, IIT Delhi, New Delhi, India

X. Mercilin Raajini
Department of ECE, Prince Shri Venkateshwara Padmavathy Engineering College, Chennai, India

G. Rajesh
Department of Information Technology, MIT Campus, Anna University, Chennai, India

Rajneesh Randhawa
Department of Computer Science, Punjabi University, Patiala, Punjab, India

Vimalathithan Rathinasabapathy
Department of ECE, Karpagam College of Engineering, Coimbatore, India

K. Martin Sagayam
Department of ECE, Karunya Institute of Technology and Sciences, Coimbatore, India

Kumud Saxena
Noida Institute of Engineering and Technology, Greater Noida, India

Nikhil Sharma
HMR Institute of Technology & Management, Delhi, India

Hitesh Singh
Noida Institute of Engineering and Technology, Greater Noida, India

Pulkit Singh
Department of Electronics and Communication Engineering, National Institute of Technology, Raipur, Chhattisgarh, India

Shweta Sinha
Amity University, Haryana, India

A. Sivasangari
School of Computing, Sathyabama Institute of Science and Technology, Chennai, India

Bhuvan Unhelkar
University of South Florida, Sarasota, FL, United States

Wireless sensor networks: Concepts, components, and challenges

1

Shweta Sinha and Priyanka Makkar
Amity University, Haryana, India

Chapter outline

1 Introduction

The power of information and communication technology coupled with the recent advancements is day-by-day making computing devices cheaper, smaller, and more mobile with power-packed capabilities. These phenomenal advancements have paved the way for the development of new generation, microlevel, low-cost sensing devices with high computational and communicational abilities for data processing. These sensing devices are called sensor nodes or motes. Each sensor node has unique capabilities that may range from simple nodes that are capable of sensing singular

Security and privacy issues in IoT devices and sensor networks. https://doi.org/10.1016/B978-0-12-821255-4.00001-8

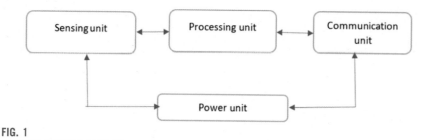

FIG. 1

Fundamental units of wireless sensor network.

physical phenomena to more complex nodes with advance capabilities of sensing signal and radio frequencies of varying data rates.

The efficiency and performance of an individual sensor node depend upon the processing units of the nodes. Each sensing nodes consist of four powerful subsystems: the sensing, the processor, the communicational, and the power subsystems. Fig. 1 presents a basic structure of a sensing node along with its fundamental units. An enormous number of these tiny nodes are integrated to create a sensor network that communicates through a wireless channel for information sharing and cooperative processing. There are a plethora of options available to decide for the topology of connection between these nodes. The basic aim of the organizational structure of the interconnection of nodes is to offer a powerful combination of computing, communication, and distributed sensing that leverage the power of the entire network even though the individual capabilities of nodes are limited. The choice of sensors is based upon the physical property to be sensed or monitored. The operational capabilities of the sensor nodes of the network make it exclusive as compared with the already existing communication systems like MANET. This uniqueness also creates some limitations to network. Some of the unique characteristics of WSN are as follows:

- Limited computation, storage, and energy capacity of the nodes.
- Self-configurable, that is, network arrangements for communication, is carried out without intervention.
- Highly unreliable network, prone to damage and malicious reading as the nodes, is placed in open and harsh environmental conditions.
- Limited power consumption is allowed as the nodes are deployed in remote areas and are battery operated.
- Scalable network makes the inclusion of new nodes into the network easier.
- Capable of handling node failure as data are redundantly stored at multiple sensors, and in case of failure, it guarantees data availability.
- A densely populated network that requires the node to be deployed very closely.

The WSNs connect the physical world with the computing world to provide a wide-ranging application relating to serious infrastructural issues. The sensors are capable of sensing sound, light, air and water qualities, and soil attributes like moisture and fertilizer requirements and also the size, weight, and dimension of objects. At present, they are being deployed at an accelerated pace. They enable a humongous amount of applications ranging from precision agriculture, environmental monitoring, healthcare, and supply-chain management, influencing the quality of life and economy. They have been an essential component of armed force communication and intelligence system and are now becoming prominent in the civilian domain also. Close surveillance of opposite forces, movement of vehicles, and monitoring the action of the forces are some of the common areas of deployment in military arrangements [1]. Monitoring of temperature and humidity that affects crops, condition of livestock [2], lighting in office spaces, and measurements of carbon dioxide emission are few of the environmental condition monitoring applications where sensors are playing an important role. WSN sensors with capabilities of audio and video sensing are being deployed for security and surveillance purpose at the airport and office complexes [3]. These sensors, when combined with the actuators, can also control the amount of water and fertilizers in the soil, cooling/heating of buildings, etc. In recent times the humungous number of sensors have been deployed in the healthcare domain. They have been playing an active role in patient monitoring and overcoming the shortage of healthcare personnel [4]. The demand for this self-configurable network is increasing day by day. There is a need for continuous efforts to overcome the limitations of the sensor network for its horizon to expand to day-to-day activities.

1.1 Network design objective

The unique characteristics of sensor nodes and varying requirement of applications pose several challenges in the implementation of WSN. Design issues related to WSN have been mentioned in [1, 5–7]. Distinct requirements give rise to differing design objectives; meeting them all is impossible. But there are a few that need to be considered by all to an extent possible in the application domain. These design objectives are as follows:

- Network cost: Cost associated with the network establishment is one of the primary design objectives to be met. As the sensor nodes are placed in open and harsh environmental conditions, they are prone to damage. A large number of nodes are needed by the applications, and these nodes can't be reused, so reducing the cost associated with each of the network nodes will reduce the overall cost. Reduction in size of the nodes will impact the power consumption of the node and will also impact its cost.
- Network coverage: The deployment of WSN is for communication purpose. Covering a larger span of the area is the objective of the network design. Increasing the coverage area will increase the demand for sensor nodes and in turn power consumption and cost. Proper selection of the network topology and data delivery model can help in enhancing the coverage while keeping the cost unchanged.

- Power consumption: The sensor nodes are battery operated and are often placed in remote and difficult areas. Their deployment makes it difficult for changing or recharging the batteries for increasing their life span. It is desirable to have the working principles of the network and the design of sensor nodes to be energy efficient and reduce the demand for power.
- Scalability: The quantum of nodes in a network can range from a few hundred to thousands conditioned upon the application for which they have been deployed. The network architecture and protocols should apply to different network sizes.
- Communication reliability: The wireless medium is more error prone and unreliable. The reliability of communication over this medium depends upon the error control and correction mechanism. WSN should have protocols for this purpose.
- Security: Often the sensor nodes are deployed in adverse areas, where they are more vulnerable to challenges. Due to constrained computing capability and limited power, resource encryption techniques are not easily applied in sensor networks. There should be some security mechanism devised for sensor networks to provide data security and prevent malicious attacks on the network.
- Utilization: Cost associated with the nodes and the communication channel impacts the network cost. The bandwidth of the communication channel is limited. Design of protocols should provide maximal utilization of the communication channel and less energy consumption by the sensor nodes during communication.
- Adaptability: The connection between nodes change dynamically, maybe due to addition, deletion, or failure of nodes. This dynamic positioning results in a change in topology along with the node distribution in the geographic area. The protocols working for WSNs should be adaptive to these changes.

1.2 Technological background

Due to the evolution of MEMS in the past few years, WSN has turned out as a new class of communication network. MEMS is a key technology for manufacturing tiny electromechanical components. By the use of different micromachining techniques that involve several fabrication processes [8, 9], these tiny components are integrated to obtain low-cost and low-powered nodes with sensing capabilities. Another key technology associated with the evolution of WSN is wireless communication. Long traditional wireless networks have extensively used the wireless medium for communication. This technology has moved through significant advances related to antenna technology, synchronization, medium access control, and routing methods. Sensor nodes are infrastructure less and communicate over lossy lines. The limited nonrenewable energy supply poses additional challenges. To overcome these challenges and maximize the life span of the network, new protocols need to be designed focusing on the efficient management of energy resources [1]. Conventional wireless systems use radio frequency (RF) for communication. But the capability of RF to provide an omnidirectional link for communication requires large radiators that are unsuitable for energy-constrained sensor nodes. The search for another possible communication medium led to the selection of free-space optical communication. The radiators used are reduced in size, and high transmission efficiency due to higher antenna gain is achieved. The overhead associated with the time, frequency, and code division multiple access was reduced due to spatial division multiple access provided by optical communication. But the need for a line of sight communication limits the use of optical communication in many applications of sensor nodes. Research is still going on to overcome these limitations [10].

Sensor network's success depends upon the available low-cost resources that include low-power hardware and easily implementable software platform. The minimal size of hardware components has been made available with the help of MEMS technology. These devices required incorporation of energy-efficient mechanism for optimization of hardware design. Several techniques are included to achieve the optimal performance of the network hardware component. The hardware component of WSN consists of low-power embedded general-purpose PCs and PDAs capable of running a lightweight operating system such as Linux. The second component sensor nodes are commercial off the shelf chips available in different variants with differing capabilities. Their choice depends upon the application to be built, and these also include a sensor that required low power and has sensing, computation, and communicational capabilities. The software component needed by WSN should be capable enough to handle files, task scheduling, memory allocation, and networking. These requirements can be fulfilled by operating system. The typical operating system for sensor network is TinyOS [11] and TinyGALS [12]. These OS are able to work on a resource-constrained hardware platform.

1.3 Network architecture

In a typical network architecture, multiple sensor nodes are densely deployed in the area of interest. These sensor nodes are capable of sensing and capturing information from the surrounding; these nodes send the sensed data to sink node in the network that is located nearby or inside the sensing region. These sinks can command or query the sensor nodes, and they may serve as a gateway in the network. The role of gateway nodes is to connect with the outside network via the Internet and transfer the process data to the user in need. The connection between the sensor nodes and the sink nodes can be a direct link, that is, single-hop connection that has to be a long-distance connection. This type of connection is inefficient as they require more energy during communication, reducing the life span of the network.

The other way for connecting the sink node and the sensor nodes is through multiple hops. This category of connection leads to short-distance communication, where the interconnection of sink and sensor node is done by several intermediate nodes using less power during transmission in short distance. This multihop connectivity can be carried out in two structures: flat architecture and

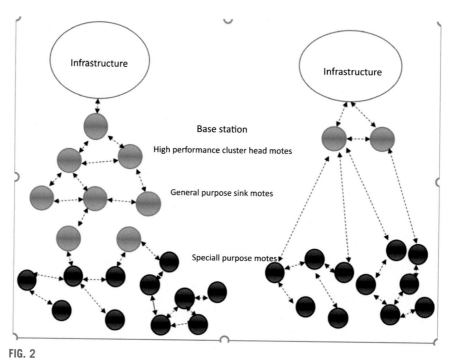

FIG. 2

Hierarchical and flat architecture of WSN.

the hierarchical architecture. Fig. 2 represents the two structures. In the *flat archi-tecture* of the WSN, all sensor nodes are of equal capability and are designated with a similar task of sensing and transmission. This architecture uses a data-centric approach, and the sink node uses flooding technique to send a query to the sensor nodes. Using this approach of data transmission, each node in the network receives the data, but only the intended node responds back to the sink.

In the case of *hierarchical architecture,* the nodes are placed into a cluster. These clusters can be formed on the basis of their capabilities and responsibilities. Each cluster has a representative called the cluster head (CH). The sensor nodes of the cluster communicate with the CH for data transfer. The CH then processes and aggregates the data received and send it to the sink node. Usually the selection of CH is guided by the energy contained in the nodes. The nodes having more energy is designated as the CH and is responsible for processing and communication in the network. Nodes with low energy are given the general task of sensing the network. The energy of the nodes decreases with time, and hence the demand for a change of CH may arise that leads to change in network structure. Undoubtedly the mecha-nism increases the energy efficiency of the network, but the selection of CH requires some efficient mechanism to conserve energy.

1.4 Classification of WSN

The deployment of WSN is application specific. Each application has different needs that are delegated to the sensor nodes of the network. As these networks are self-configurable, to complete their designated task, they work differently for each application. Based on their working criteria, WSN can be classified as follows:

- *Deterministic and nondeterministic network:* A preplanned deployment of sensor nodes leads to the deterministic WSN, and these rarely exist. The nondeterministic network has randomly deployed nodes and support flexibility and scalability of the network.
- *Single sink and multiple sink network:* Network with single sink node receives data from all the nodes in the vicinity at only one point, thus increasing the load at one point. The multiple sink network provides load balancing by allowing multiple nodes to work as a sink. The sensing nodes can share the data to the one that is at the smallest distance from there.
- *Single-hop and multihop network:* Single-hop communication requires long-distance transmission that leads to more energy consumption, whereas multihop uses intermediate nodes for transmission of data from sensing nodes to sink and consumes less energy.
- *Homogeneous network and heterogeneous network:* In a homogeneous network, each node has equal capability related to storage, processing, and energy and is assigned with similar responsibilities. The nodes in the heterogeneous network differ in their characteristics. One node may have more power than the others and can be designated with additional responsibilities. This arrangement increases the network life span.

2 WSN communication pattern

One of the desirable properties of WSN is to support mobile applications and hence require the flexible deployment of nodes that communicate over the wireless link [13, 14]. The sensor network's major activity involves communication among the nodes for data transfer. To increase the life span of the network, all the communication across the nodes must be done in an energy-efficient way, and a need to follow some predefined rules generates the requirement of a protocol stack for WSN.

2.1 Protocol stack of WSN

The layered architecture of WSN protocol is designed to overcome the challenges of wireless communication. This protocol stack integrates data and network protocols and promotes the cooperative effect of sensor nodes [1]. Fig. 3 represents the protocol stack of WSN. The protocol stack consists of the physical layer, data link layer, network layer, transport layer, and application layer. Each layer is further divided

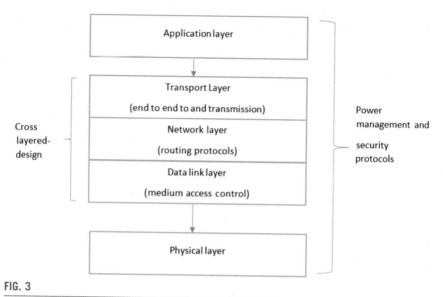

FIG. 3

Layered structure of WSN protocols.

into management planes responsible for power management of sensor nodes by implementing efficient power management mechanism at different layers in the protocol stack. These management layers are responsible for task management that handles task distribution across the sensors and management of configuration and reconfiguration of nodes in the network.

The physical layer of the protocol stack takes care of the modulation, frequency selection, and encryption of the message transmitted in the communication network and provides service to the data link layer. The data link layer is responsible for energy-efficient communication with proper bandwidth utilization. Error control and flow control on the shared network have to be managed by data link layer protocols. The efficient routing of data from source to destination is one of the key factors in the direction of achieving efficient, long-lived network. The network layer handles this responsibility by defining a set of rules for data delivery. Apart from energy-efficient communication, reliability is the most desirable characteristics of WSN that is needed by users. Reliability refers to the correct source to sink transmission, and the transport layer protocol works for this purpose. TCP and UDP protocol work for connectivity between source to sink and sink to the user. The application layer is responsible for providing application-specific software that translates data based upon requirements.

Every layer of WSN is vulnerable to several attacks. These attacks are done either in active or in passive mode [15]. Since layers of the network all have different protocols, they exhibit different security vulnerabilities. Table 1 presents some of the prominent threats at these network layers along with the solutions to overcome them.

Table 1 Threats and their counter measure at each WSN layer.

Network layer	Attacks	Threat	Security approach
Physical layer	Jamming	A denial of service attack where adversary disrupts the network or part of it by interfering with communication radio frequency of SNs	Priority messages, spread spectrum techniques, lower duty cycle
	Sybil attack	Adversary can steal or fabricate the identity of any authentic node and pose accordingly	Count on total number of sensor nodes
	Exhaustion	Repeated collision during communication leads to energy wastage	Rate limitation
Link layer	Collision	Collision even in small portion can lead to checksum error and can demand retransmission	Error correcting code
	Unfairness	A denial of service attack where to get advantage the attacker degrades the quality of service that may lead to missing transmission deadline	Small frames
	Black hole attack	False advertising of zero cost route influences the routing paths selected by protocols	Use of mobile agent, trust-based mechanism
Network layer	Hello flood attack	Adversary with high-power captures a sensor node and broadcast hello messages in communication channel and declare itself as neighbor to many distant nodes	Authentication, bidirectional link
Transport layer	Misdirection	When the packets pass through malicious nodes, they are forwarded to false direction which makes the packets unreachable	Move to sleep mode
	Flooding	A denial of service attack where continuous transmission of connection request leads to flooding in the network	Fixed upper limit on connections
	Session hijack	Attacker spoofs the victim's IP address and sends its own with that identity	SSL/TLS authentication
Application layer	Malicious code attack	Infects the OS and user applications, spread themselves to the network and cause slowdown	Cryptographic approach, authentication and encryption

Sybil attack is one of the major threats to WSN. It may originate at physical layer and influence link and network layer. The main purpose of this attack is to target fault-tolerant schemes such as multipath and distributed storage. Encryption and authentication techniques can be employed to tackle it to some extent.

2.2 Medium access control at data link layer

Shared communication medium in any network needs some mechanism to control and resolve the access issues. Reliability and efficiency of WSN also depend on this mechanism. WSN structure requires data propagation from the collecting/sensor nodes to the sink/base stations where it can be worked upon further. Due to shared communication, medium collision is one of the major concerns in any wireless network, and so is in WSN. Retransmission of messages has to be avoided to preserve the energy of the nodes. To resolve the issues associated with the collision in shared medium, medium access control (MAC) protocol should work toward arbitrating access to the shared channel and avoid data collision. It is desirable that the operation of an access protocol be distributed in nature.

Traditional MAC protocols are divided into two categories: contention-based MAC protocols and contention-free MAC protocols. In contention-based MAC the nodes sharing the common medium are required to contend for the transmission channel. During contention a collision may occur, and the MAC protocols define a mechanism to resolve it. ALOHA, CSMA, and CSMA/CA are contention-based protocols. In contention-free MAC protocols, the shared communication channel is divided into several subchannels based on frequency, time, or code. These subchannels are allocated to nodes one at a time to avoid collision. TDMA, FDMA, and CDMA are typical examples of traditional contention-free protocols. To handle peculiarities associated with WSN such as the dense deployment of nodes, severe constraints on power consumption, storage limitations, and higher unreliability traditional MAC protocols have to be modified accordingly. The desirable property of the WSN protocols is to reduce energy consumption in the network. Apart from this, latency and security are equally important features desirable in the protocols. Design issues associated with MAC protocols are guided by issues such as energy efficiency, required to provide/maximize lifetime of individual sensor nodes; scalability, scalable to varying network size; adaptability, can be easily modified to changing node density and topology; latency and throughput, should be able to provide maximum transfer of data with minimal delay; and channel utilization and fairness, efficient use of bandwidth by providing equal opportunity to each node for using the channel. It's impossible to have a single protocol that can resolve all the issues discussed earlier, but efforts have been put forward to resolve them to a maximum extent, giving priority to reduced energy consumption. Collision, overhearing of channel, idle listening of channel, and control overhead are a few factors that need to be taken care to reduce the consumed energy [16].

The MAC protocols for WSN can be categorized as contention-based and schedule-based protocols. The schedule-based protocols (TRAMA [17], UNPF [18]) follow an approach of creating an assignment that every node of the network has to follow. These are further categorized as synchronous, asynchronous, and hybrid schedule-based protocols. The synchronous category of protocols requires topological information to know about its node and also need to time synchronize to schedule the transfer of data when both are active. The asynchronous protocols do not need this information, and communication takes place in different active cycles.

The contention-based protocols used back off mechanism in case of collision and avoided the overhead associated with synchronization. Ye et al. [19, 20] propounded a sensor-MAC (S-MAC) protocol that mainly focuses on energy efficiency, collision avoidance, and scalability. While achieving these, it compromises with fairness and latency. To achieve the best performance and these issues, this protocol implements periodic listen and sleep mechanism to handle the *idle-listen* problem. To avoid overhearing the protocol uses control packets and sends the node to sleep. Collision issue is managed by sensing virtual and physical carriers. Lu et al. [21] proposed a protocol for data gathering in the sensor network. This protocol, D-MAC, is energy efficient with low latency. It provides solution for multihop path and employs wake-up schedule to enable continuous data forwarding. P-MAC [18] is a WSN protocol that differs from most of the MAC protocols in the sleeping–listening pattern of nodes. Several other protocols have stable sleeping–listening pattern, but P-MAC determines this timing based on traffic loads in neighborhood. wiseMAC [22] is an energy-efficient protocol that works for multihop and infrastructure networks. This protocol uses nonpersistent CSMA and preamble sampling. CSMA-based MAC protocol [23] is energy efficient and achieves fair bandwidth allocation for every type of network traffic. It works for multihop network and combines CSMA with adaptive rate control mechanism to control the network traffic.

Apart from this, progressive signaling mechanism is used to control the rate of origin of data. But the main drawback is associated with high control overhead and hidden terminal problem experienced by this protocol. Overall, it can be observed that contention-based protocols have multiple benefits like low complexity in implementation and ability to handle varying traffic patterns, whereas a schedule-based algorithm like TDMA has the intrinsic capability of energy conservation as there is no collision, so no overhead and retransmission are required.

2.3 Network layer

Several challenges are faced during the design of protocols for WSN, and energy efficiency is the major challenge associated with WSN. The energy consumed by sensors in sensing and processing is negligible as compared with the energy consumed during communication. Two different approaches exist in sampling data from WSN: the push and the pull approach. The push approach continuously samples the data at a fixed interval and pushes it into the network creating a bottleneck at sink node but reduces the delay associated with query response. In the pull approach the sampling is done on demand by sink. This approach reduces the energy consumption but increases the number of message transmission in the network giving rise to latency in query response.

The data delivery in WSN can be modeled based on the perspective of application traffic or on the perspective of infrastructure-based communication [24]. The application perspective classifies the delivery model as periodic, query driven, event driven, or hybrid [24]. In the periodic approach, data from nodes to the sink are transferred in a fixed interval. The query-driven approach transfers the data in response to the query raised by the user after matching the criteria. In the event-driven approach, the transfer from node to sink is triggered by any event that meets some predefined threshold values, and in the hybrid approach, one or more models are combined together.

Infrastructure-based perspective classifies the flow pattern of data as unicast, broadcast, multicast, and geocast. On-to-one communication between the nodes or node and sink is referred to as unicast. Broadcast is one-to-many communication, and multicast refers to group communication. It follows a broadcast pattern for a specific group. When this transmission is associated with some specific geographic location, it is termed as geocast [25]. Fig. 4 represents the data delivery model of WSN. The protocol design, especially at the network layer, must take into consideration all these data delivery models.

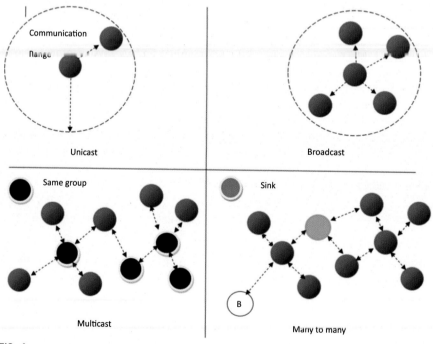

FIG. 4

Infrastructure-based data delivery model of WSN.

2.3.1 Routing protocol

Due to multiple numbers of node interconnection, a decision regarding the route for propagation has to be taken at each intermediate station. The efficiency of the computed route also impacts the sensor network performance. Several approaches appear for classification of routing protocols. No addressing scheme like IP addressing exist in WSN so that location information can be utilized for identification. The rules for the relay of data in WSN are often governed by the geographic location (coordinate values) of the nodes in the network. Depending on the location and placements of nodes, the communication can be either long or short distance, which influences the energy consumed. The protocols using this approach are termed as location-based protocols. Considering the communication entities involved in routing protocols, they can be classified as node centric or data centric. The node-centric protocol uses the node address for data routing between the nodes, whereas attribute-based addressing is used in data-centric routing. Another way to classify WSN is based on network architecture: the flat or hierarchical structure of the network. The communication route can be single hop or multihop depending upon the network architecture. Robustness of network is one of the desirable properties of WSN.

Due to the highly unreliable nature of sensor nodes, mechanism is needed to provide measures to cope up with network failures. Existence of multiple paths between the nodes can handle the link failure in the network. The protocols can be classified based on path redundancy among the nodes. The network can be static or dynamic. Static networks are the one where both the sink and the sensor node are immobile and fixed at their location. Handling routes for communication in this setup is easy. But in dynamic network the nodes may be mobile and sink static or vice versa. Change in their location influences the route between the two, and hence this frequent change has to be taken care by routing protocol. The protocols can thus also be classified considering the network dynamics. Quality of service (QoS) can again be treated as classification measure. The sensing application varies in nature, and they may have different requirements in terms of reliability, security, and robustness. A routing protocol may not be able to handle all of these collectively, and they have to set trade-off between them. Not all sensor nodes of any network have same life span and capability and neither the similar responsibility. Their responsibilities and capabilities directly influence energy consumption. Routing protocols must consider these factors also. Fig. 5 represents the classification of routing protocols and also list of protocols of each category.

Each of these protocols possesses some merits and has a few limitations associated with them. Here in Table 2, we outline the merits and demerits of a few protocols in each category. More details on others can be obtained from [7, 50].

2.4 Transport layer

The role of the transport layer in the network is to provide end-to-end data delivery along with a congestion control mechanism. The data delivery can be reliable or unreliable service, whereas the congestion control mechanism can be implemented

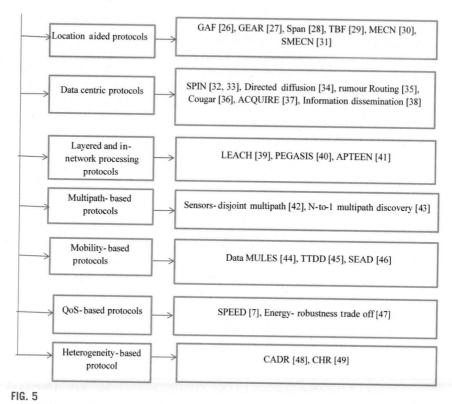

FIG. 5

Routing protocols and their classification [26–51].

in centralized or distributed fashion depending upon the application requirements. For providing a reliable service, mechanism of retransmission of the lost packet has to be done without delay. This can be achieved by assigning a sequence number to the packets and acknowledging the receipt of the packet at the destination. The protocols at this layer also take care of the data relay rate of each node in the network and work to avoid a bottleneck in the transmission channel. FLUSH [51] and RMST [52] are few reliable protocols at the transport layer that also have distributed congestion control mechanism.

3 WSN design challenges

Undoubtedly, WSNs depict the latest trends of Moore's law with miniature ubiquitous computing devices. Although packed with massive power, WSN faces numerous challenges due to their peculiarities. Some of the major challenges are outlined here as follows:

Table 2 Routing protocols with their merits and limitations.

Category	Name of protocol	Merits	Limitations
Location-aided protocols	Geographic Adaptive Fidelity (GAF)	Energy coservation	Processing overhead
	Geographic and Energy Aware Routing (GEAR)	Use recursive data dissemination in small region instead of flooding	Lifetime depletion
	Span	Interfacing or power saving	Single failure point
	Small Minimum Energy Communication Network (SMECN)	Energy-efficient path	Huge number of broadcast message
Data-centric protocol	Sensor Protocols for Information via Negotiation (SPIN)	Energy aware and resource aware	Does not apply to mobile links, works only for lossless network
	Directed diffusion	Improved robustness and scalability of network	Event handling and user interaction high
	Rumor routing	Efficient node failure handling, tunable for a range of query/event ratios	Optimal parameters dependent on topology, does not guarantee delivery
	Cougar	In-network data aggregation saves energy efficiently	Requires extra storage and synchronization
Layered and in-network	LEACH	Efficient resource utilization, uniform load distribution	Energy dissipation, does not guarantee good CH distribution
	PEGASIS	Increase network lifetime, reduce overhead by dynamic clustering minimization of bandwidth usage	Requires global information of network, unsuitable for variable network topology
	APTEEN	Works for hybrid network, flexible	Additional Overhead and complexity associated with threshold
Multipath based	Disjoint paths	Low scalability, failure on one path leads to selection of alternate paths	More resilient to sensor failures, less energy efficient

Continued

Table 2 Routing protocols with their merits and limitations—cont'd

Category	Name of protocol	Merits	Limitations
Mobility based	Two-Tier Data Dissemination Model (TTDD)	Scalable, handles multiple mobile sinks	Cannot handle mobile sensor nodes
QoS based	SPEED	Congestion Avoidance, less energy consumption	
Heterogeneity based	Constrained anisotropic diffusion routing (CADR)	Information gain, minimized latency	

- *Energy-efficient communication:* The sensors in the network are low-powered units. Having very less amount of energy reserved, some mechanism has to be applied to elongate the lifetime and obtain better communication performance.
- *Extended network coverage:* Besides efficient communication, one of the major concerns is to completely cover the area under concern and get more accurate results from the sensed data. The placement of sensor nodes is very dense, but it should be placed such that optimum coverage is provided by each of them.
- *Limited resources:* The resource required by a sensor network can be a software module or any hardware component. The sensor node has restricted processing power and storage capability; due to these limitations, they can provide restricted functionality. These hardware limitations impose constraints on software performance and topology design. It is a challenge to maintain the computational and communication latency to the minimal with these limited resources.
- *High unreliability:* The sensor nodes are generally placed in an open area and are exposed to harsh environmental conditions. Apart from this the placement of nodes are not always preplanned, and most often, no careful engineering or planning is possible before placing the nodes. Sometimes, they are airdropped in remote areas or are placed manually. These conditions make them highly prone to damage, and they can often sense malicious data making the information unreliable.
- *Load balancing:* The node density of WSN is very high. It is desirable to obtain the optimum usage of these nodes. The node responsibility should be assigned in a manner to maintain a balance between communication, processing, and sensing the responsibility of each node. A mechanism such as dynamic cluster formation and cluster head rotation mechanism can balance the lifetime of each node.
- *Diverse applications:* WSN is nowadays implemented for several diverse application areas, ranging from ocean-based WSN to a simple application like pressure and temperature monitoring. Different application has different needs, and fulfilling these diverse requirements using a single protocol is a challenge.

4 Comparative analysis of optimized clustering algorithm

Energy consumption is one of the stringent requirements of WSN. A considerable amount of work has been done for efficiently managing and utilizing the energy stored in WSN nodes [53, 54]. One way to reduce energy consumption is to implement listen and sleep mechanism. It has been observed that the hierarchical network structure performs better than other network structures in terms of energy consumption and network lifetime [53]. The hierarchical network is based on the arrangement of sensor nodes in clusters, with each cluster having a representative: the cluster head (CH). This CH is responsible for gathering data from all the nodes in the cluster and further sending the aggregated data to the base station or sink. The communication among CHs is often done in multihop scenario. The CH to sink communication is single hop; hence the distance between sink and CH is a considerable parameter in CH selection.

A further saving of energy can be achieved by listening and sleep rule inside each cluster. The effective formation of cluster and selection of CH influence the energy consumption of the network. The choice of CH is influenced by several factors or node attributes. The best candidate for being elected as CH is the node that has maximum leftover energy, surrounded by a maximum number of nodes, and is situated at a minimal distance from the sink [54]. To obtain an optimal network performance in terms of energy utilization or consumption, the selection of cluster head is dynamic, and the nodes are given this responsibility in rotation. Apart from CH selection, there is also a provision for selection of second node termed as vice cluster head to work as a support to CH and help increase lifetime [55].

4.1 Cluster formation scenario

Initially a random cluster is formed from the given data and based on the energy stored; a node from each cluster is randomly chosen and elected as CH. Fig. 6 represents the flow sequence of the task involved in the creation of a cluster and cluster head.

Once the clusters are created, the cluster head sends advertisement message to all the nodes in the network. For each node the distance between the node and the CH is computed using Euclidean distance measure. The nodes are placed in the cluster with minimal distance. If the distance between the sink and the sensor node is less than the distance between CH and node, then the nodes directly communicate with the sink. When the nodes join any cluster, it sends a CH-join message that lets the other nodes and the head know about its presence. The CH assigns time slot for collecting data from each node. Once the data have been collected, it is aggregated by the CH and further transmitted to sink. During this complete process the nodes may go to sleep, but CH has to continuously live. This reduces the energy of CH; also with time, few nodes die living a sparse network after some time. The clusters are restructured, and CHs are selected in each cycle.

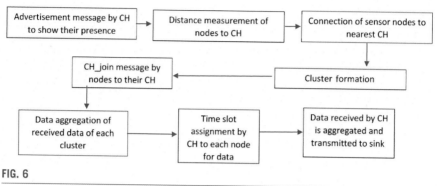

FIG. 6

Flow graph for cluster formation and cluster head creation.

There are several traditional algorithms for creation and selection of CH and also restructuring as, and when required, all mainly uses probabilistic techniques [39]. Clustering in WSN being an NP-hard problem, these probabilistic approach-based methods are not able to optimize the cluster creation. The advancements in the field of computational intelligence (CI) and machine learning (ML) have influenced the clustering paradigm, and optimization of clustering algorithm has been proposed using CI and ML. Research [56, 57] has highlighted that the optimized clustering techniques have performed better than the traditional approaches in terms of energy consumption, fault tolerance, scalability, and reliability.

4.2 Optimized clustering strategy

The clustering approach under this category is also termed as intelligent clustering methods and is based on developments in the field of ML and CI. The availability of ML/CI approach for clustering in sensor networks is classified as fuzzy logic, neural network, genetic algorithm, and swarm intelligence [56, 57]. Here, we utilize some of the intelligently optimized clustering algorithms and compare them based upon five parameters:

- data delivery rate, that is, number of received data by base or other nodes as compared with sent data by nodes;
- scalability that helps in identifying the performance of the algorithm with an increased number of nodes;
- consumption of energy that helps in identifying the lifetime of the network;
- data aggregation that helps in identifying the role of cluster head and helps in knowing the network lifetime;
- analysis of algorithm by identifying its role as a multihop or direct communicating algorithm help in obtaining the energy consumption and lifetime of the network.

5 Evaluation of clustering methods for optimization

In this section, we evaluate few clustering algorithms that are based upon CI/ML approaches by simulating them in a large WSN. The algorithms selected for evaluation are a few selected from fuzzy logic, neural networks, genetic algorithm, and swarm intelligence. We present a brief summary of the selected algorithms and evaluate their performance for the five parameters in the next section.

- *Fuzzy logic techniques (FL):* Fuzzy logic is based upon the approximate reasoning that measures the uncertainty of data using imprecise values. Two algorithms based on this approach are evaluated for its performance.

CHEF [58]—The cluster head formation utilizes the information like distance to the base station and the energy stored in the nodes.
   LEACH-FL [59]—CH is selected based upon the power of the node, distance from sink, and density of node.

- *Genetic algorithm techniques (GA):* This approach is influenced by genetic evolution in living elements. It uses an adaptive heuristic approach for optimization. The algorithms evaluated here are as follows:

LEACH-GA [60]—Works in a centralized manner and uses the information such as node status, Ids, and node location sent to sink. GA operations are performed on them for optimization.
   GABEEC [61]—It's an energy-efficient technique that uses GA. It initially creates a static cluster, but the cluster head is dynamic, and based upon the residual energy, the nodes inside the clusters are selected as CH.

- *Neural network approach (NN):* This approach is inspired by biological neurons and is represented as layers of interconnection between the neurons and automatically learns the relationship among them. Only a few works have utilized this approach for clustering:

SOM-based clustering [62]—Self-organizing map strategy is utilized for cluster formation based on minimum distance from CH. The CH-dead message is sent directly to the sink station when the leftover energy of the nodes becomes less than some predefined threshold value. Selection of the new head is again done using SOM strategy.

- *Swarm intelligence techniques (SI):* SI approach is influenced by the collective behavior of social insects or animal societies. They are further classified as particle swarm optimization (PSO), ant colony optimization (ACO), and bee colony optimization (BCO) [57, 63–65]. The recent trend in intelligent optimization is influenced by these strategies and uses one of PSO, ACO, and BCO.

PSO-C [66]—This algorithm is based on particle swarm intelligence, an evolutionary technique for computation influenced by bird flocking or swarm theory. This is a population-based small algorithm that starts with a random population solution. PSO-C uses a centralized approach and takes into account the distance between the cluster head and other nodes of the cluster along with the remaining energy of the CH candidate. A threshold value based on average energy is used for candidate selection.

EBAB [67]—It is energy-balanced, ACO-based technique. ACO uses the ants' behavior of communicating with group members. The technique utilizes the meta-heuristic approach and models the problem as a search problem. The problem is represented as a path graph, and the minimal path is obtained. EBAB produces variable size clusters with the ones near the sink being smaller. The overall cluster formation and cluster head creation are done by broadcasting messages and acknowledgments from two sides.

5.1 Modeling of system

To evaluate the performance of the algorithm simulation approach is obeyed. The system model is described here, and the measurements obtained are compiled to present the comparative analysis of these algorithms.

- Huge number of nodes placed in a 150 × 150 square region.
- The node distribution is uniform throughout the area.
- Initially the nodes dissipate their own information by sending "Hello" message to all the nodes in the network.
- At the start the cluster size is based on all the deployed nodes, and the number of clusters is kept at optimal. As the nodes die, reclustering is performed, and, in this process, the number of clusters may merge to create fewer clusters.
- It is assumed that the sink of the network is a node without any energy limitations and enhanced computational capabilities. They are placed in the center of the region of interest.
- The consumption of energy for sending m bit data is proportional to the square of distance d between the nodes. In the case of long distance, amplification is required, which also consumes energy. Energy is needed during the reception of data and is independent of the distance d.

6 Result and analysis

The performance of the algorithms has been evaluated in a simulated environment, and the results are compiled. Simulation parameters used for the comparison of clustering algorithms is presented in Table 3.

The performance of the system was measured with uniform node deployment in the square region. Experiments were done by varying the number of nodes in the

Table 3 Parameters used during simulation.

Parameters	Value	Parameters	Value
Network span	150 × 150	Number of nodes	100
Node distribution	Uniform throughout the area	Node energy coefficient (a_i)	Random number between (0,10)
Node initial essential energy	0.5 J–1 J	E_{elec}	50 nJ/bit
BS position	(75,75)	E_{frs}	10 nJ/bit/m^2
Data packet size	1000 bytes	E_{amp}	0.0013 pJ/bit/m^4
Rounds	4000	E_{DA}	5 nJ/bit/signal

where E_{elec} is energy consumed per bit when emitter components are running,
E_{frs} is energy consumed by the unit power amplifier in the free space, and
E_{amp} is energy consumed by the unit power amplifier in double path propagation model.

defined region, that is, sparse and dense distribution of nodes (number of nodes varying between 100 and 200). The comparison is obtained based on the average performance of these techniques. Table 4 represents the comparison of the algorithm.

The comparative analysis highlights that in terms of energy consumption, the SI-based techniques provide the best result. In general, centralized processing-based algorithms consume less energy as compared with the distributed approaches. Also, in terms of data delivery, it has been observed that the data delivery rate for SI techniques was better in a medium-scale network and these techniques are suitable for application design using the amount of data transfer needed but only when the node deployment is average. In terms of scalability, most of the system provides low scalability, which is desirable for clustering. But it is mainly application-specific, so looking into other factors, we say that ML/CI approaches are useful. Also, it is observed that the SI approach is more scalable as compared with others, the reason being its nature to work in collaboration with large population. Figs. 7 and 8

Table 4 Comparison of optimized clustering methods.

Algorithm	ML/CI	Data delivery rate	Energy consumption	Data aggregation	Scalability	Multihop
CHEF	FL	–	High	No	Low	No
LEACH-FL	FL	Average	High	Yes	Low	No
LEACH-GA	GA	Low	High	Yes	Low	No
GABEEC	GA	–	Low	Yes	Low	–
SOM based	NN	Average	Average	Yes	Low	–
PSO-C	PSO	Average	Average	Yes	Low	No
EBAB	ACO	Low	Average	–	High	Yes

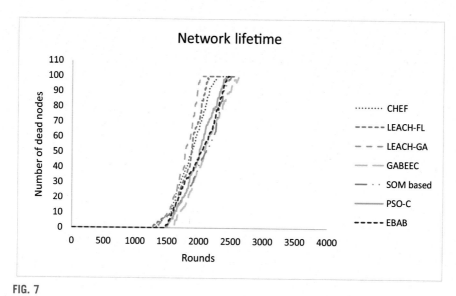

FIG. 7

Comparison of network lifetime of seven algorithms.

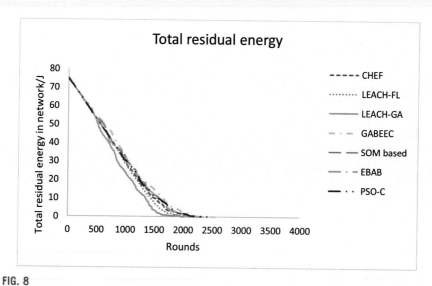

FIG. 8

Comparison of system energy consumption.

represent the network lifetime measured in terms of number of dead nodes in different rounds of simulation of each protocol and the overall energy of the network in different rounds, respectively. The cumulative energy is again an indirect measure of network life span and also highlights the energy conservation aspect of each algorithm.

7 Conclusion and future work

WSN has unique characteristics, and hence the concept, protocols, and algorithms of a conventional system cannot be directly extended to this domain. Identification of those differences is essential for development in the domain of WSN. This chapter discusses the fundamental concepts of WSN. In WSN, challenges related to deployment, scalability, and coverage are of foremost concern. The chapter highlights those challenges and reviews the approaches for dealing with them. Medium access protocol influences network performance and energy consumption. The content presented in this chapter focuses on the fundamental concepts of the MAC layer, and a thorough review is outlined to identify the major challenges in this layer. WSN is mainly a communication network, and energy consumption and security are the vital aspects to control the network lifetime, performance, and reliability. A discussion on routing protocol and data dissemination has been presented in this chapter with an aim to provide a better understanding of the role of routing protocols. Also the security threats during the course of communication have been outlined. The chapter discusses some of the CI/ML techniques for cluster optimization and compares their performance in a simulated environment.

Advancements of technology in communication have overcome several issues associated with WSN performance. However, there are still several open challenges that require attention.

i. Due to the unguided medium of communication, wireless communication has always been vulnerable to massive security attacks. Energy efficiency, security, and QoS are some of the major issues that need attention.

ii. Clustering is one of the major techniques used for energy consumption and data dissemination. ML-based methods have shown to outperform traditional methods for energy conservation. More efforts are required for building hierarchical clustering principles. The need is to enhance the ML-based protocols further to consider asymmetric link problems, node failure, and mobility issues as a priority.

iii. Sensor nodes are being utilized by the military and critical civilian applications. The need is to work more on the security aspects of WSN. One of the major requirements is to enhance security performance and counterattacks while optimizing reliability and throughput.

References

[1] I.F. Akyildiz, W. Su, Y. Sankarasubramaniam, E. Cayirci, Wireless sensor networks: a survey, Comput. Netw. 38 (4) (2002) 393–422.

[2] A. Mainwaring, J. Polastre, R. Szewczyk, D. Culler, J. Anderson, Wireless sensor networks for habitat monitoring, in: Proceedings of 1st ACM International Workshop on Wireless Sensor Networks and Applications (WSNA' 02), Atlanta, GA, September, 2002, 88–97.

[3] F. Zhao, L. Guibas, Wireless Sensor Networks: An Information Processing Approach, Morgan Kaufmann Publishers, San Francisco, CA, 2004.

[4] R. Jafari, A. Encarnacao, A. Zahoory, F. Dabiri, H. Noshadi, M. Sarrafzadeh, Wireless sensor networks for health monitoring, in: Proceedings of 2nd Annual International Conference on Mobile and Ubiquitous Systems: Networking and Services (MobiQuitous' 05), July, 2005, 479–481.

[5] J. Pan, Y. Hou, L. Cai, Y. Shi, S.X. Shen, Topology control for wireless sensor networks, in: Proc. 9th ACM Int. Conf. on Mobile Computing and Networking, San Diego, USA, September, 2003, 286–299.

[6] O. Younis, S. Fahmy, HEED: a hybrid, energy-efficient, distributed clustering approach for ad hoc sensor networks, IEEE Trans. Mobile Comput. 3 (4) (2004) 366–379.

[7] K. Akkaya, M. Younis, A survey on routing protocols for wireless sensor networks, Ad Hoc Netw. 3 (3) (2005) 325–349.

[8] R.F. Pierret, Introduction to Microelectronic Fabrication, Addison-Wesley, Menlo Park, CA, 1990.

[9] S.D. Senturia, Microsystem Design, Kluwer Academic Publishers, Norwell, MA, 2001.

[10] L. Ma, H. Leung, D. Li, Hybrid tdma/cdma mac protocol for wireless sensor networks, J. Netw. 9 (10) (2014) 2665.

[11] J. Hill, R. Szewczyk, A. Woo, D. Culler, S. Hollar, K. Pister, System architecture directions for networked sensors, in: Proceedings of 9th International Conference on Architectural Support for Programming Languages and Operating Systems (ASPLOS IX), Cambridge, MA, November, 2000, 93–104.

[12] E. Cheong, J. Liebman, J. Liu, F. Zhao, TinyGALS: a programming model or event-driven embedded systems, in: Proceedings of 18th Annual ACM Symposium on Applied Computing (SAC' 03), Melbourne, FL, March, 2003, 698–704.

[13] G. Elhayatmy, N. Dey, A.S. Ashour, Internet of things based wireless body area network in healthcare, in: Internet of Things and Big Data Analytics toward Next-Generation Intelligence, Springer, Cham, 2018, 3–20.

[14] N. Dey, A.S. Ashour, F. Shi, S.J. Fong, R.S. Sherratt, Developing residential wireless sensor networks for ECG healthcare monitoring, IEEE Trans. Consum. Electron. 63 (4) (2017) 442–449.

[15] B. Bhushan, G. Sahoo, Recent advances in attacks, technical challenges, vulnerabilities and their countermeasures in wireless sensor networks, Wirel. Pers. Commun. 98 (2) (2018) 2037–2077.

[16] W. Ye, J. Heidemann, D. Estrin, An energy-efficient MAC protocol for wireless sensor networks, in: INFOCOM 2002, 21st Annual Joint Conference of IEEE Computer and Communications Societies, New York, 2002.

[17] V. Rajendran, K. Obraczka, J. Garcia-Luna-Aceves, Energy-efficient collision-free medium access control for wireless sensor networks, in: ACM Intl. Conference on Embedded Networked Sensor Systems (SenSys), Los Angeles, CA, November, 2003.

[18] J. Ding, K.M. Sivalingam, R. Kashyapa, L.J. Chuan, A multi-layered architecture and protocols for large-scale wireless sensor networks, in: Proc. IEEE Semiannual Vehicular Technology Conference-Fall, Orlando, FL, October, 2003.

[19] W. Ye, J. Heidemann, D. Estrin, An energy efficient MAC protocol for wireless sensor networks, in: Proceedings of IEEE INFOCOM' 02, New York, NY, June, 2002, 1567–1576.

[20] W. Ye, J. Heidemann, D. Estrin, Medium access control with coordinated adaptive sleeping for wireless sensor networks, IEEE/ACM Trans. Netw. 12 (3) (2004) 493–506.

[21] G. Lu, B. Krishnamachari, C.S. Raghavendra, An adaptive energy - efficient and low - latency MAC for data gathering in wireless sensor networks, in: Proceedings of 18th International Parallel and Distributed Processing Symposium (IPDPS' 04), Santa Fe, NM, April, 2004, 224–231.

[22] G.-S. Ahn, E. Milazzo, A.T. Campbell, S.G. Hong, F. Cuomo, Funneling-MAC: a localized, sink - oriented MAC for boosting fidelity in sensor networks, in: Proceedings of ACM Conference on Embedded Networked Sensor Networks (SenSys'06), Boulder, Colorado, November, 2006, 293–306.

[23] J.M. Khan, R.H. Katz, K. Pister, Next century challenges: mobile networking for smart dust, in: Proceedings of 5th International Conference on Mobile Computing and Networking (MobiCom' 99), Seattle, WA, August, 1999, 271–278.

[24] S. Tilak, N.B. Abu-Ghazaleh, W. Heinzelman, A taxonomy of wireless micro-sensor network models, ACM SIGMOBILE Mob. Comput. Commun. Rev. 6 (2) (2002) 28–36.

[25] L. Villalba, A. Orozco, A. Cabrera, C. Abbas, Routing protocols in wireless sensor networks, Sensors 9 (11) (2009) 8399–8421.

[26] Y. Xu, J. Heidemann, D. Estrin, Geography - informed energy conservation for ad hoc routing, in: Proceedings ACM/IEEE MobiCom' 01, Rome, Italy, July, 2001, 70–84.

[27] Y. Yu, R. Govindan, D. Estrin, Geographical and Energy-Aware Routing: A Recursive Data Dissemination Protocol for Wireless Sensor Networks, Technical Report UCLA/CSD-TR-01-0023, UCLA Computer Science Department, 2001.

[28] B. Chen, K. Jamieson, H. Balakrishnan, R. Morris, Span: an energy-efficient coordination algorithm for topology maintenance in ad hoc wireless networks, in: Proceedings ACM MobiCom' 01, Rome, Italy, July, 2001, 85–96.

[29] B. Nath, D. Niculescu, Routing on a curve, ACM SIGCOMM Comput. Commun. Rev. 33 (1) (2003) 155–160.

[30] V. Rodoplu, T.H. Meng, Minimum energy mobile wireless networks, IEEE J. Sel. Areas Commun. 17 (8) (1999) 1333–1344.

[31] L. Li, J.Y. Halpern, Minimum - energy mobile wireless networks revisited, in: Proceedings IEEE ICC' 01, Helsinki, Finland, June, 2001, 278–283.

[32] W.R. Heinzelman, J. Kulik, H. Balakrishnan, Adaptive protocols for information dissemination in wireless sensor networks, in: Proceedings ACM MobiCom' 99, Seattle, WA, August, 1999, 174–185.

[33] J. Kulik, W. Heinzelman, H. Balakrishnan, Negotiation-based protocols for disseminating information in wireless sensor networks, Wirel. Netw 8 (2/3) (2002) 169–185.

[34] C. Intanagonwiwat, R. Govindan, D. Estrin, J. Heidemann, F. Silva, Directed diffusion for wireless sensor networking, IEEE/ACM Trans. Netw. 11 (1) (2003) 2–16.

[35] D. Braginsky, D. Estrin, Rumor routing algorithm in sensor networks, in: Proceedings ACM WSNA, in Conjunction with ACM MobiCom' 02, Atlanta, GA, September, 2002, 22–31.

[36] Y. Yao, J. Gehrke, The cougar approach to in-network query processing in sensor networks SGIMOD, Theatr. Rec. 31 (3) (2002) 9–18.

[37] N. Sadagopan, B. Krishnamachari, A. Helmy, The ACQUIRE mechanism for efficient querying in sensor networks, in: Proceedings SNPA' 03, Anchorage, AK, May, 2003, 149–155.

[38] D. Ganesan, R. Govindan, S. Shenker, D. Estrin, Highly-resilient, energy-efficient multipath routing in wireless sensor networks, in: Proceedings ACMMobiHoc' 01, Long Beach, CA, October, 2001, 251–254.

[39] W. Heinzelman, A. Chandrakasan, H. Balakrishnan, An application-specific protocol architecture for wireless microsensor networks, IEEE Trans. Wirel. Commun. 1 (4) (2002) 660–670.

[40] S. Lindsey, C.S. Raghavendra, PEGASIS: power-efficient gathering in sensor information systems, in: Proceedings IEEE Aerospace Conference, vol. 3, Big Sky, MT, March, 2002, 1125–1130.

[41] A. Manjeshwar, D.P. Agrawal, APTEEN: a hybrid protocol for efficient routing and comprehensive information retrieval in wireless sensor networks, in: Proceedings IPDPS' 01, San Francisco, CA, April, 2001, 2009–2015.

[42] D. Ganesan, R. Govindan, S. Shenker, D. Estrin, Highly-resilient, energy-efficient multipath routing in wireless sensor networks, Mobile Comput. Commun. Rev. 5 (4) (2001) 10–24.

[43] W. Lou, An efficient N - to −1 multipath routing protocol in wireless sensor networks, in: Proceedings IEEE MASS' 05, Washington, DC, November, 2005, 1–8.

[44] R.C. Shah, S. Roy, S. Jain, W. Brunette, Data MULEs: modeling a three-tier architecture for sparse sensor networks, in: Proceedings SNPA' 03, Anchorage, AK, May, 2003, 30–41.

[45] F. Ye, H. Luo, J. Cheng, S. Lu, L. Zhang, A two-tier data dissemination model for large-scale wireless sensor networks, in: Proceedings ACM/IEEE MobiCom' 02, Atlanta, GA, September, 2002, 148–159.

[46] H.S. Kim, T. Abdelzaher, W.H. Kwon, Minimum–asynchronous energy dissemination to mobile sinks in wireless sensor networks, in: Proceedings ACM SenSys' 03, Los Angeles, CA, November, 2003, 193–204.

[47] B. Krishnamachari, Y. Mourtada, S. Wicker, The energy-robustness tradeoff for routing in wireless sensor networks, in: Proceedings IEEE ICC' 03, Seattle, WA, May, 2003, 1833–1837.

[48] M. Chu, H. Haussecker, F. Zhao, Scalable information-driven sensor querying and routing for ad hoc heterogeneous sensor networks, Int. J. High-Performance Comput. Appl. 16 (3) (2002) 293–313.

[49] X. Du, F. Lin, Improving routing in sensor networks with heterogeneous sensor nodes, in: Proceedings IEEE VTC' 05, Dallas, TX, September, 2005, 2528–2532.

[50] B. Bhushan, G. Sahoo, A comprehensive survey of secure and energy efficient routing protocols and data collection approaches in wireless sensor networks, in: Proceedings ICSPC, IEEE, 2017, 294–299.

[51] S. Kim, R. Fonseca, P. Dutta, A. Tavakoli, D. Culler, P. Levis, S. Shenker, I. Stoica, Flush: a reliable bulk transport protocol for multihop wireless networks, in: Proceedings of the 5th International Conference on Embedded Networked Sensor Systems, ACM, 2007, 351 365.

[52] N. Rani Das, R.R. Sahoo, D.M. Sar, An analytical survey of various congestion control and avoidance algorithms in WSN, Int. J. Emerg. Trends Technol. Comput. Sci. 2 (4) (2013).

[53] J.Y. Yu, P.H.J. Chong, A survey of clustering schemes for mobile ad hoc networks, IEEE Commun. Surv. Tutorials 7 (1) (2005) 32–48.

[54] A.A. Abbasi, M. Younis, A survey on clustering algorithms for wireless sensor networks, Comput. Commun. 30 (14–15) (2007) 2826–2841.

[55] A. Mehmood, J. Lloret, M. Noman, H. Song, Improvement of the wireless sensor network lifetime using LEACH with vice-cluster head, Ad Hoc Sens. Wirel. Netw. 28 (2015) 1–17.

[56] S. Sirsikar, K. Wankhede, Comparison of clustering algorithms to design new clustering approach, Procedia Comput. Sci. 49 (2015) 147–154.

[57] B. Solaiman, A. Sheta, Computational intelligence for wireless sensor networks: applications and clustering algorithms, Int. J. Comput. Appl. 73 (2013) 1–8.

[58] J.S. Kim, Y. Park, T.C. Han, CHEF: cluster head election mechanism using fuzzy logic in wireless sensor networks, Adv. Commun. Technol. ICACT 654–659 (2008) 17–20.

[59] G. Ran, H. Zhang, G. Shulan, Improving on LEACH protocol of wireless sensor using fuzzy logic, Inf. Comput. Sci. 3 (2010) 767–775.

[60] J. Liu, C. Ravishankar, V LEACH-GA: genetic algorithm-based energy-efficient adaptive clustering protocol for wireless sensor networks, Int. J. Mach. Learn. Comput. 1 (2011) 79–85.

[61] S. Bayrakli, S.Z. Erdogan, Genetic algorithm based energy efficient clusters (GABEEC) in wireless sensor networks, Procedia Comput. Sci. 10 (2012) 247–254.

[62] M. Cordina, C.J. Debono, Increasing wireless sensor network lifetime through the application of SOM neural networks, in: Proceedings of the 3rd International Symposium on Communications, Control and Signal Process, Rhodes, Greece, 12–14 March, 2008.

[63] R.V. Kulkarni, A. Förster, G.K. Venayagamoorthy, Computational intelligence in wireless sensor networks: a survey, IEEE Commun. Surv. Tutor. 13 (2011) 68–96.

[64] M. Bhanderi, H. Shah, Machine learning for wireless sensor network: a review, challenges and applications, Adv. Electron. Electr. Eng. 4 (2014) 475–486.

[65] A. Arya, A. Malik, R. Garg, Reinforcement learning based routing protocols in WSNs: a survey, Int. J. Comput. Sci. Eng. Technol. 4 (2013) 1401–1404.

[66] N.M.A. Latiff, C.C. Tsimenidis, B.S. Sharif, U. Kingdom, Energy-aware clustering for wireless sensor networks using particle swarm optimization, in: Proceedings of the 18th Annual IEEE International Symposium on Personal, Indoor and Mobile Radio Communications, Montreal, QC, Canada, 3–7 September, 2007.

[67] L. Wang, R. Zhang, A.N. Model, An energy-balanced ant-based routing protocol for wireless sensor networks, in: Proceedings of the 5th International Conference on Wireless Communications, Networking and Mobile Computing, Beijing, China, 24–26 September, 2009.

Secure performance of emerging wireless sensor networks relying nonorthogonal multiple access

2

Dinh-Thuan Do[a], Anh-Tu Le[b], and Narayan C. Debnath[c]

[a]Wireless Communications Research Group, Faculty of Electrical and Electronics Engineering, Ton Duc Thang University, Ho Chi Minh City, Vietnam, [b]Faculty of Electronics Technology, Industrial University of Ho Chi Minh City, Ho Chi Minh City, Vietnam, [c]School of Computing and Information Technology, Eastern International University, Thu Dau Mot City, Vietnam

Chapter outline

1 Brief history of IoT communications related to multiple access technique

The Internet of Things (IoT) is emerging technique providing large number of applications and services through the Internet medium. IoT and its applications are available to satisfy increasing demands from users in wireless communications. It is necessary to introduce securely and intelligently exchange data technique in such IoT environment of dense deployment of IoT devices such as actuators, cheaper sensor nodes, and smart devices. Since industry, health care, financial businesses, and

29

Security and privacy issues in IoT devices and sensor networks. https://doi.org/10.1016/B978-0-12-821255-4.00002-X

robust device authentication are secure communication via the connectivity of any IoT device, data confidentiality and privacy characterizations need be addressed [1, 2].

In the context of massive connections, a large number of communication devices are served in a complex platform of the IoT, e.g., controllers, sensors, and actuators. However, a vital and challenging task is required to exhibit ubiquitous connectivity, which is necessary in deployment of IoT-based communication systems. The orthogonal multiple access (OMA) is known as popular multiple access scheme applied in current IoT networks.

Typically, unlicensed spectrum is allocated to wireless devices in IoT network, and these bands are suitable to design popular applications. The emerging techniques introduced in IoT include LoRa, Sigfox, and Weightless to provide wide range of services from basic to complex. For instance, vehicle-to-everything (V2X) communications or health monitoring tools are implementations of IoT in practice. The existing IoT platforms need improvement to achieve enhanced capability, low latency, and high data rates. This book chapter aims at the introduction of multiple access without orthogonal characterization; it is so-called nonorthogonal multiple access (NOMA) [3–6]. The implementation of NOMA in various applications of 5G still adopts secure method, which enhances security at physical layer [3]. NOMA benefits spectral efficiency improvement and other advantages reported in [5]. It can be achieved low latency and high energy efficiency as employing NOMA [6] (Table 1).

Motivated by these analyses, we consider secure performance of IoT system relying NOMA. Main contributions of this book chapter are summarized as follows:

- In this book chapter, various scenarios of NOMA deployment are considered to achieve improved bandwidth efficiency. In particular, this chapter presents

Table 1 Access schemes in IoT wireless.

		Technologies	
Parameters	**LoRa, SigFox**	**Machine-type communications (mMTC), narrow band-IoT**	**5G-New Radio (5G-NR)**
Spectrum access	Unlicensed (ISM bands)	Licensed	Licensed
Multiple access technique	OMA	OMA	OMA and NOMA
Maximum range (km)	50	25	2
Throughput	600 (bps)	1 (Mbps)	>100 (Mbps)
Channel bandwidth	15 (kHz)	200 (kHz)	5–400 (MHz)

theoretical analysis of emerging networks relying NOMA. For example, NOMA can be proposed to employ in cognitive radio network, cooperative network, and extra consideration on physical layer security for such NOMA-assisted IoT system.

- Regarding physical layer security, it can be further introduced how IoT system secures main signal transmission to sensors using NOMA. Such system is then beneficial from methods of physical layer security (PLS). When PLS and NOMA are combined, these promising techniques benefit spectral efficiency, energy efficiency, and improved secure performance. However, the source node in IoT system needs to adjust power allocation fractions to serve each user who need separate demands at downlink side in IoT system.
- The outage probability and secrecy outage probability (SOP) in exact form are examined to show different NOMA users' performance. A strict agreement can be seen as comparing analytic results and results achieved from Monte-Carlo simulations, and hence these illustrations are examined in simulation section.

The structure of this book chapter is as follows: Section 2 presents the system model and fundamental of NOMA in IoT system. Section 3 provides the context of cooperative network applied in IoT system. Then, Section 4 considers cognitive radio-assisted IoT system. Secure performance at physical layer is presented in Section 5. In Section 6, we conduct a performance evaluation of IoT system, and, finally, in Section 7, we conclude and discuss possible future directions.

2 Basic fundamentals of NOMA

However, bandwidth in wireless networks is limited, while ubiquitous connections from sensors in IoT applications results in higher challenge. Spectrum efficiency benefits to IoT systems as in [7]. The time/frequency/code resources can be divided to users in NOMA, and hence massive users are allowed to access upon their demand with different quality of service (QoS) [8].

In principle, cooperative diversity [9] is architecture that implements dedicated relays to make multiple paths to provide links between source and destination. Such technique is relevant transmission to combat fading. The authors in [9] indicated that amplify-and-forward (AF) relay protocols can obtain the diversity order as two. NOMA has attracted much attention in cooperative systems in recent years [10]. In [9, 10], authors presented cooperative NOMA system in which users have higher priority to achieve information thanks to better channel acquired. In particular, strong users play an important role as relay to forward information to weak users. Recently, main advantage of employments of cooperative relaying technique in NOMA is the improvement of the spectral efficiency [10,11] .

3 NOMA and application in cooperation network

In cooperative network using NOMA, it is necessary to consider a system model that contains the source node S that serves two sensors (destinations), D1 and D2. This IoT system is able to serve massive sensors, and such two sensors are representative users that need be studied. However, links between the source and the sensors are too far and it needs the help of group of relays, and relaying links are considered to provide higher reliability concern. It is noted that the superimposed signal $\sqrt{P_s a_1} x_1 + \sqrt{P_s a_2} x_2$ is sent in the first phase to both NOMA destination users via relaying node R. Here, a_1, a_2 are power allocation coefficients to the first sensor and the second sensor, while x_1, x_2 are signals that are targeting these sensors. We call h_{SR}, h_{RD1}, h_{RD2} as channel corresponding link S-R, R-D1, and R-D2. P_s, P_R transmit power at the source and the relay. Then the received signal at the relay is given as

$$y_R = h_{SR}\left(\sqrt{P_s a_1} x_1 + \sqrt{P_s a_2} x_2\right) + n_R \tag{1}$$

where n_R denote as the additive Gaussian noise with zero mean and variance of N_0.

It is noted that a variable gain to amplify the received signal in Amplify-and-Forward (AF) relaying mode and it is performed during the second time slot. The gain factor is given by

$$\psi^2 = \frac{P_R}{P_s|h_{SR}|^2 + N_0} \tag{2}$$

It leads to the ability to compute the received signal at D1, and it is performed in the second phase and is given by

$$\begin{aligned} y_{D_1} &= \psi h_{SR} h_{RD_1}\left(\sqrt{a_1} x_1 + \sqrt{a_2} x_2\right) + h_{RD_1}\psi n_R + n_{D_1} \\ &= \psi h_{SR} h_{RD_1}\left(\sqrt{a_1} x_1 + \sqrt{a_2} x_2\right) + h_{RD_1}\psi n_R + n_{D_1} \end{aligned} \tag{3}$$

4 NOMA and cognitive radio-assisted IoT system

The improved spectral efficiency resulted from implementing NOMA together with cognitive radio (CR) to form CR-assisted system [12, 13]. In such system the licensed primary users (PU) with higher quality of service (QoS) has higher priority to be served, while opportunistic services are available to the unlicensed secondary users (SU). As main constraint the SU has constrained transmit power. In the instantaneous signal-to-interference-plus-noise ratio (SINR) of the PU, SU needs to be computed before achieving other performance evaluations, because the same spectrum shared to both the PU and SU in same time and hence higher spectral efficiency is benefited from implementation of CR-assisted system.

4.1 System model of IoT relying NOMA and CR

In this cognitive radio-inspired NOMA network, it is necessary to consider the performance of secondary network (SN). Such SN contains the source node S that serves two sensor nodes D1 and D2. However, links between the source and the sensors are too far and it needs the help of group of relays, and both direct link and relay link are considered. It is assumed that selected relay R_k needs to serve far sensor nodes. In the first phase the source node S sends the superimposed signal $\sqrt{P_s a_1} x_1 + \sqrt{P_s a_2} x_2$ to both NOMA destination users and R_k. We call h_{SD1}, h_{SD2}, h_{SRk} as channel corresponding link S-D1, S-D2, and S-selected relay. P_s, P_{Rk} transmit power at source and relay. Then, at direct link, the received signal at the two destination users D1 and D2 is given, respectively, as (Fig. 1)

$$y_{D_1} = h_{SD_1} \left(\sqrt{P_s a_1} x_1 + \sqrt{P_s a_2} x_2 \right) + n_{D_1} \tag{4}$$

$$y_{D_2} = h_{SD_2} \left(\sqrt{P_s a_1} x_1 + \sqrt{P_s a_2} x_2 \right) + n_{D_2} \tag{5}$$

where n_{D1}, n_{D2} denote as the additive Gaussian noise with zero mean and variance of N_0. Similarly, the received signal at relay is given by

$$y_{R_k} = (\sqrt{a_1} x_1 + \sqrt{a_2} x_2) h_{SR_k} + n_{R_k} \tag{6}$$

Using a variable gain to amplify the received signal during the second time slot, R_k has an explicit factor as

$$\psi^2 = \frac{P_{R_k}}{P_s |h_{SR_k}|^2 + N_0} \tag{7}$$

The second phase needs to process at user D1, and such received signal is formulated by

$$
\begin{aligned}
y_{D_1} &= \left(\psi y_{R_k} + \eta_{R_k D_1} \right) h_{R_k D_1} + n_{D_1} \\
&= \psi h_{SR_k} h_{R_k D_1} (\sqrt{a_1} x_1 + \sqrt{a_2} x_2) + h_{R_k D_1} \psi n_{R_k} + n_{D_1} \\
&= \psi h_{SR_k} h_{R_k D_1} (\sqrt{a_1} x_1 + \sqrt{a_2} x_2) + h_{R_k D_1} \psi n_{R_k} + n_{D_1}
\end{aligned}
\tag{8}
$$

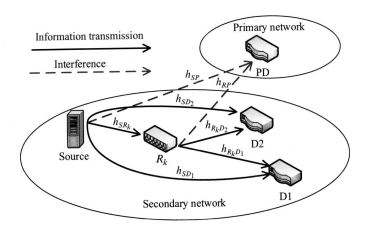

FIG. 1

CR-NOMA applied in network with multiple relays.

Similarly, it can be expressed the signal at user D2 as

$$y_{D_2} = \psi h_{SR_k} h_{R_k D_2} (\sqrt{a_1} x_1 + \sqrt{a_2} x_2) + h_{R_k D_2} \psi n_{R_k} + n_{D_2} \tag{9}$$

It is required that maximum tolerable interference power Q make decision on other interference measured at P. Then, P_S and P_{R_k} are controlled by such power. Particularly, $E\{|h_{SP} x_S|^2\} \le Q$ and $E\{|h_{R_k P}| G y_{R_k} + \eta_{R_k D_2}\} \le Q$. As a result, it is denoted that $P_S = Q/|h_{SP}|^2$ and $P_{R_k} = Q/|h_{RP}|^2$.

The signal to interference plus noise ratio (SINR) is first important metric using to achieve further metrics. In particular, SINR to detect D1's signal is formulated as

$$\gamma_{SD_1} = \frac{\rho a_1 |h_{SD_1}|^2}{\rho a_2 |h_{SD_1}|^2 + W} \tag{10}$$

in which $W = |h_{SP}|^2$. Then the SINR can be obtained at D2 as

$$\gamma_{SD_{12}} = \frac{\rho a_1 |h_{SD_2}|^2}{\rho a_2 |h_{SD_2}|^2 + W} \tag{11}$$

Using successive interference cancellation (SIC) in NOMA, signal-to-noise ratio (SNR) can be expressed at D2:

$$\gamma_{SD_2} = \frac{\rho a_2 |h_{SD_2}|^2}{W} \tag{12}$$

We $Z_k = |h_{R_k P}|^2$, $\gamma_{SR_k} = \frac{\rho |h_{SR_k}|^2}{W}$, $\gamma_{R_k D_i} = \frac{\rho |h_{R_k D_i}|^2}{Z_k}$, $\rho = \frac{Q}{N_0}$. As relaying link is activated, the received signal-to-noise ratio (SNR) at D1 can be computed by

$$\gamma_{kD_1} = \frac{\gamma_{SR_k} \gamma_{R_k D_1} a_1}{\gamma_{SR_k} \gamma_{R_k D_1} a_2 + \gamma_{SR_k} + \gamma_{R_k D_1} + 1} \tag{13}$$

Based on operation of relaying link, SINR needs be computed to detect signal of D1 before decoding signal of D2 as

$$\gamma_{kD_{12}} = \frac{\gamma_{SR_k} \gamma_{R_k D_2} a_1}{\gamma_{SR_k} \gamma_{R_k D_2} a_2 + \gamma_{SR_k} + \gamma_{R_k D_2} + 1} \tag{14}$$

Then, it can be obtained SINR at user D2 using SIC, and it is expressed by

$$\gamma_{kD_2} = \frac{\gamma_{SR_k} \gamma_{R_k D_2} a_2}{\gamma_{SR_k} + \gamma_{R_k D_2} + 1} \tag{15}$$

4.2 Outage probability analysis in case of partial relay selection
4.2.1 Performance analysis of the first user

Considering Partial Relay Selection (PRS) scheme applied to selected relay [12], the index of selected relay (best relay) is denoted by $k^* = \arg \max_{k \in \{1,\dots,K\}} \{\gamma_{SR_k}\}$. Hence, at a given threshold γ_{th1}, it can be formulated the outage probability (OP) as

$$OP_1^{PRS}(\gamma_{th1}) = \Pr\left(\max\{\gamma_{SD_1}, \gamma_{k^* D_1}\} < \gamma_{th1}\right) \tag{16}$$

It happens in two cases to exhibit the closed-form expression of outage probability. Detailed computation of such performance can be explained as.

Case I ($\gamma_{th1} < \frac{a_1}{(a_2 + \alpha)}$): Outage performance of the first user OP_1^{PRS1} is expressed by

$$OP_1^{PRS1} \approx 1 - \left(1 + \frac{\gamma_{th1}\Omega_{SP}}{\Omega_{SD_1}\theta}\right)^{-1} - \left(1 + \frac{\gamma_{th1}\Omega_{RP}}{\Omega_{RD_1}\vartheta}\right)^{-1}$$

$$\times K \sum_0^{K-1} \binom{K-1}{n} \frac{(-1)^n \gamma_{th1}\Omega_{SP}}{(n+1)\Omega_{SD_1}\theta} \times \left(1 + \frac{(n+1)\gamma_{th1}\Omega_{SP}}{\Omega_{SR}\vartheta}\right)^{-1} \tag{17}$$

$$\times \left(1 + \frac{\gamma_{th1}\Omega_{SP}}{\Omega_{SD_1}\theta} + \frac{(n+1)\gamma_{th1}\Omega_{SP}}{\Omega_{SR}\vartheta}\right)^{-1}$$

Case II ($\gamma_{th1} > \frac{a_1}{(a_2 + \alpha)}$): It is simpler way to compute OP_1^{PRS2}, and it is given as

$$OP_1^{PRS2} \approx 1 - \left(1 + \frac{\gamma_{th1}\Omega_{SP}}{\Omega_{SD_1}\theta}\right)^{-1} \tag{18}$$

where $\theta = \rho(a_1 - \gamma_{th1}a_2)$ and $\vartheta = \rho(a_1 - \gamma_{th}(a_2 + \alpha))$.

4.2.2 Outage probability at D2

Such outage of the second sensor in this case is given by

$$OP_2^{PRS}(\gamma_{th2}) = \Pr\left(\max\{\gamma_{SD_2}, \gamma_{k\ D_2}\} < \gamma_{th2}\right) \tag{19}$$

Case I ($\gamma_{th2} < \frac{a_2}{\alpha}$): Such outage event can formulate OP_2^{PRS1} as

$$OP_2^{PRS} \approx 1 - \left(1 + \frac{\gamma_{th2}\Omega_{SP}}{\Omega_{SD_2}\rho a_2}\right)^{-1} - \left(1 + \frac{\gamma_{th2}\Omega_{RP}}{\Omega_{RD}\rho(a_2 - \gamma_{th2}\alpha)}\right)^{-1}$$

$$\times K \sum_0^{K-1} \binom{K-1}{n} \frac{(-1)^n \gamma_{th2}\Omega_{SP}}{(n+1)\Omega_{SD_2}\rho a_2} \times \left(1 + \frac{(n+1)\gamma_{th2}\Omega_{SP}}{\Omega_{SR}\rho(a_2 - \gamma_{th2}\alpha)}\right)^{-1} \tag{20}$$

$$\times \left(1 + \frac{\gamma_{th2}\Omega_{SP}}{\Omega_{SD_2}\rho a_2} + \frac{(n+1)\gamma_{th2}\Omega_{SP}}{\Omega_{SR}\rho(a_2 - \gamma_{th2}\alpha)}\right)^{-1}$$

Case II ($\gamma_{th2} > \frac{a_2}{\alpha}$): Such outage event can formulate OP_2^{PRS2} as

$$OP_2^{PRS} \approx 1 - \left(1 + \frac{\gamma_{th2}\Omega_{SP}}{\Omega_{SD_2}\rho a_2}\right)^{-1} \tag{21}$$

Remark The outage probability analysis in the case of Opportunistic Relay Selection (ORS) can be achieved by similar analysis as PRS [12].

5 Improving security at physical layer

To guarantee secure transmission the security is high priority to implement IoT network under impacts from surrounding eavesdroppers. In conventional encryption techniques, it recovers the secret messages to prevent eavesdroppers [14]. However,

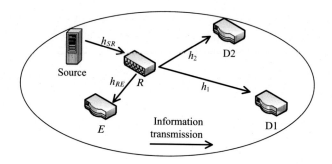

FIG. 2

Secure performance of IoT system relying NOMA.

the traditional technique, such as encryption-based techniques, leads to difficulties and vulnerabilities regarding secret key management. Fortunately the eavesdropper and the legitimate receiver exhibit distinguished signal by allying new technique, namely, physical layer security (PLS). In addition, PLS techniques exploit physical signal to detect vulnerability to keep reliable signal transmission [15, 16] (Fig. 2).

Consider that IoT system is relying on NOMA scheme. Such system includes a base station (S) which is requires to serve, two users (D1, poor user, and D2, strong user). The transmission link between the BS and far users need the help of a relay. Unfortunately, normal transmission affords impacts from cellular network containing an eavesdropper (E). In this scenario the BS is placed in the center of the serving area, while the cell-edge area includes both far users and user E. Therefore no direct links exist at hops between the far users and user E with the BS. These nodes are facilitated by single antenna. It is assumed that all the channels follow independent Rayleigh fading. More importantly, user E is able to overhear the information from relay to users. In particular, overhearing information includes the forwarding signal from the relay and decoding signal at users.

It is necessary to consider the SNR at relay to detect signal x_1, and it can be expressed as

$$\gamma_{r,1}^{AF} = \frac{|h_1|^2 |h_{SR}|^2 a_1}{|h_1|^2 |h_{SR}|^2 a_2 + \frac{1}{\rho}|h_1|^2 + \frac{1}{\rho}|h_{SR}|^2 + \frac{1}{\rho^2}} \tag{22}$$

After performing SIC, SNR is at relay to detect x_2

$$\gamma_{r,2}^{AF} = \frac{\rho |h_{SR}|^2 |h_2|^2 a_2}{|h_2|^2 + |h_{SR}|^2 + \frac{1}{\rho}} \tag{23}$$

Similarly, SNR computations are achieved for eavesdropper E. We deploy AF relaying mode to serve signal forwarding to user D1 and D2; rates at these nodes can be written as

$$C_{r,i}^{AF} = 0.5 \log_2\left(1 + \gamma_{r,i}^{AF}\right), i = 1, 2 \tag{24}$$

Similarly, we formulate rate at user E as

$$C_{E,i}^{AF} = 0.5 \log_2 \left(1 + \gamma_{E,i}^{AF} \right), \tag{25}$$

It is noted that the average SNR of the illegal link is $\gamma_{E,i}^{AF} = a_i \rho |h_{SE}|^2$.

In principle, at for user D1 and D2, the secrecy rate of the NOMA-based IoT system is given by

$$C_i^{AF} = \left[C_{r,i}^{AF} - C_{E,i}^{AF} \right]^+, \tag{26}$$

where $[x]^+ = \max \{x, 0\}$.

Therefore, when C_1^{AF} or C_2^{AF} is less than threshold target rates, outage behavior occurs. The secure outage probability (SOP) is written as [17,18].

$$SOP^{AF} = \Pr \left(C_1^{AF} < R_1 \, or \, C_2^{AF} < R_2 \right)$$

$$= 1 - \Pr \left(\frac{1 + \gamma_{r,1}^{AF}}{1 + \gamma_{E,1}^{AF}} > C_{th}^1, \frac{1 + \gamma_{r,2}^{AF}}{1 + \gamma_{E,2}^{AF}} > C_{th}^2 \right) \tag{27}$$

$$= 1 - \text{Pr1},$$

where $C_{th}^i = 2^{2R_i}$ and denoting the target data rates for user D_i, $i = 1, 2$ as R_i.

Fortunately, at high SNR regime, it can adopted the following upper bounds $\gamma_{r,1}^{AF} < \frac{a_1}{a_2}$, $\gamma_{r,2}^{AF} < \frac{\rho \rho |h_{SR}|^2 |h_2|^2 a_2}{\rho |h_2|^2 + \rho |h_{SR}|^2}$, and $\gamma_{E,i}^{AF} = a_i \rho |h_{SR}|^2$. Then an upper bound of Pr1 can be formulated as

$$\text{Pr1} < \Pr \left(\frac{1 + \frac{a_1}{a_2}}{1 + a_1 \rho |h_{SR}|^2} < C_{th}^1, \frac{\rho \rho |h_{SR}|^2 |h_2|^2 a_2}{\rho |h_2|^2 + \rho |h_{SR}|^2} > C_{th}^2 a_2 \rho |h_{SR}|^2 + C_{th}^2 - 1 \right), \tag{28}$$

where $\xi_1 = \frac{1 - a_2 C_{th}^1}{a_1 a_2 C_{th}^1}$ and $\alpha_1^2 > \alpha_2^2 (C_{th}^1 - 1)$ due to $\xi_1 = \frac{a_1 - a_2 (C_{th}^1 - 1)}{a_1 a_2 C_{th}^1} > 0$. Otherwise, such SOP equals to 1. At high SNR, approximate SOP can be rewritten as

$$\Pr \left(\min \left\{ |h_r|^2, |g_{r,2}|^2 \right\} > C_{th}^2 |h_r|^2 + B \right)$$

$$- \Pr \left(\rho |h_r|^2 > \xi_1, \min \left\{ |h_r|^2, |g_{r,2}|^2 \right\} > C_{th}^2 |h_r|^2 + B \right) \tag{29}$$

$$= P_1 - P_2$$

The following results can be achieved [17].

$$P_1 = \Pr \left(|h_{SR}|^2 > C_{th}^2 |h_{SR}|^2 + B, |h_2|^2 > C_{th}^2 |h_{SR}|^2 + B \right)$$

$$= \Pr \left(|h_{SR}|^2 > \frac{B}{1 - C_{th}^2} \right) \int_0^\infty \Pr \left(|h_2|^2 > C_{th}^2 x + B \right) f_{|h_{SR}|^2}(x) dx \tag{30}$$

$$= e^{-(\Omega_{SR}) \frac{B}{1 - C_{th}^2}} \int_0^\infty e^{-\Omega_2 \left(C_{th}^2 x + B \right)} \Omega_{SR} e^{-\Omega_{SR} x} dx = e^{-\frac{B \Omega_{SR}}{1 - C_{th}^2}} \frac{e^{-\Omega_2 B}}{\Omega_2 C_{th}^2 + \Omega_{SR}}$$

where channels $|h_1|^2$, $|h_2|^2$ are with gains Ω_1, Ω_2 for the Rayleigh channel parameters. Furthermore, it can be obtained P_2 as

$$P_2 = \Pr\left(\rho|h_{SR}|^2 > \xi_1, \min\left\{|h_{SR}|^2, |h_2|^2\right\} > C_{th}^2|h_{SR}|^2 + B\right)$$

$$P_2 = \Pr\left(\rho|h_{SR}|^2 > \xi_1\right)P_1 = e^{-\left(\frac{\Omega_{SR}\xi_1}{\rho}\right)}P_1 \tag{31}$$

Finally the SOP is most important metric to consider IoT systems, and it is expressed by

$$SOP^{AF} = 1 - P_1 + P_2 \tag{32}$$

6 Validating achievable expressions of outage behavior and secure performance via numerical simulation

Our proposed IoT system using NOMA is examined by numerical simulation to shows performance of two sensors (users). The first important metric, so-called outage probability, needs be illustrated. We set $d_{SR} = d_{RD2} = d_{RP} = 5$, $d_{RD1} = 10$ to perform simulations, and power allocation factors are $a_1 = 0.8$, $a_2 = 0.2$, and the threshold SNR $\gamma_{th1} = \gamma_{th2} = 3$.

Figs. 3 and 4 plot the outage probability performance since transmit SNR at the source is varied. By increasing transmit SNR, better outage performance can be obtained. Two users D1 and D2 show performance gap. In this regard, two relay selection schemes are applied, and two lines of outage behavior resulted by power allocation values. It can be observed that more relay provides better performance, i.e., $K = 2$, is better choice compared with case with $K = 1$. It is confirmed that the analytical curves match with asymptotic lines very well as observations in Figs. 3 and 4 and it happens at high SNR.

Fig. 5 shows the SOP curves for different parameters. It can be seen clearly that the SOP with specific SNR exhibits constant value. These SOP curves depend on target rates. Here, we set $\rho_E = 0 (dB)$. The lower curve is with $R_1 = 0.1$, $R_2 = 0.5$, while the upper cure corresponds to $R_1 = 0.5$, $R_2 = 1$. It is concluded that reliable communication if reasonable target rate and SNR of the eavesdropper link is limited.

7 Conclusion

In this book chapter, we analyzed the outage performance and secrecy performance of cooperative NOMA-based IoT systems. Main result indicated that better outage performance can be achieved at high SNR at the source. The performance gaps among two destinations happen since allocating different power allocation factors

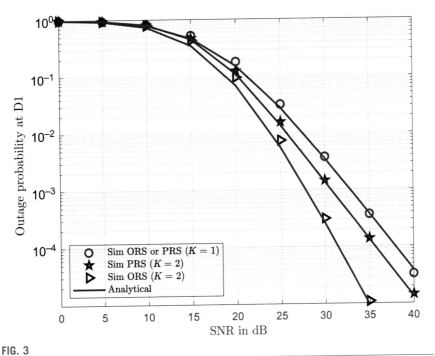

FIG. 3

Outage behavior of two relay selection modes at D1 versus SNR.

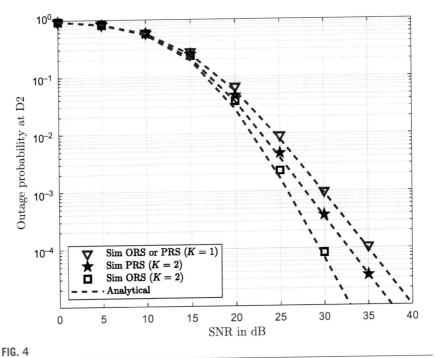

FIG. 4

Two relay section modes in evaluating outage behavior at D2.

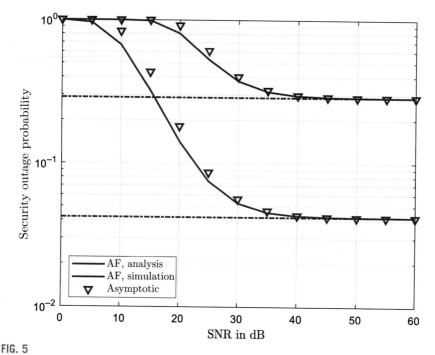

FIG. 5

Secure performance for such IoT system versus SNR.

to users. Interestingly the SOP of NOMA-based IoT systems keeps constant at high SNR. Moreover, higher spectrum efficiency can be achieved as combining NOMA and CR to implement IoT system. Furthermore, signal in IoT systems related to physical layer satisfies standard in term of secure communication by limiting performance of eavesdropper.

References

[1] J. Singh, T. Pasquier, J.M. Bacon, H. Ko, D. Eyers, Twenty security considerations for cloud-supported internet of things, IEEE Internet Things J. 3 (3) (2016) 269–284.

[2] J. Li, M. Wen, T. Zhang, Group-based authentication and key agreement with dynamic policy updating for MTC in LTE-A networks, IEEE Internet Things J. 3 (3) (2016) 408–417.

[3] D.-T. Do, M.-S. Van Nguyen, T.-A. Hoang, M. Voznak, NOMA-assisted multiple access scheme for IoT deployment: relay selection model and secrecy performance improvement, Sensors 19 (3) (2019) 736.

[4] D.-T. Do, M. Vaezi, T.-L. Nguyen, Wireless powered cooperative relaying using NOMA with imperfect CSI, in: Proc. of IEEE Globecom Workshops (GC Wkshps), Abu Dhabi, UAE, 2018, pp. 1–6.

[5] D.-T. Do, A.-T. Le, C.-B. Le, B.M. Lee, On exact outage and throughput performance of cognitive radio based non-orthogonal multiple access networks with and without D2D link, Sensors 19 (3314) (2019).

[6] D.-T. Do, M.-S. Van Nguyen, Device-to-device transmission modes in NOMA network with and without wireless power transfer, Comput. Commun. 139 (2019) 67–77.

[7] Z. Ding, Y. Liu, J. Choi, Q. Sun, M. Elkashlan, I. Chih-Lin, H.V. Poor, Application of non-orthogonal multiple access in LTE and 5G networks, IEEE Commun. Mag. 55 (2) (2017) 185–191.

[8] M. Zeng, A. Yadav, O.A. Dobre, G.I. Tsiropoulos, H.V. Poor, Capacity comparison between MIMO-NOMA and MIMO-OMA with multiple users in a cluster, IEEE J. Sel. Areas Commun. 35 (10) (2017) 2413–2424.

[9] D.-T. Do, C.-B. Le, Application of NOMA in wireless system with wireless power transfer scheme: outage and ergodic capacity performance analysis, Sensors 18 (10) (2018) 3501.

[10] T.-L. Nguyen, D.-T. Do, Exploiting impacts of Intercell interference on SWIPT-assisted non-orthogonal multiple access, Wirel. Commun. Mob. Comput. 2018 (2018) 2525492 12 Pages.

[11] T.-L. Nguyen, D.-T. Do, Power allocation schemes for wireless powered NOMA systems with imperfect CSI: system model and performance analysis, Int. J. Commun. Syst. 31 (15) (2018) e3789.

[12] D.-T. Do, A.-T. Le, NOMA based cognitive relaying: transceiver hardware impairments, relay selection policies and outage performance comparison, Comput. Commun. 146 (2019) 144–154.

[13] D.-T. Do, A.-T. Le, B.M. Lee, On performance analysis of underlay cognitive radio-aware hybrid OMA/NOMA networks with imperfect CSI, Electronics 8 (7) (2019) 819.

[14] S.L. Keoh, S.S. Kumar, H. Tschofenig, Securing the internet of things: a standardization perspective, IEEE Internet Things J. 1 (3) (2014) 265–275.

[15] X. Hu, P. Mu, B. Wang, Z. Li, On the secrecy rate maximization with uncoordinated cooperative jamming by single-antenna helpers, IEEE Trans. Veh. Technol. 66 (5) (2017) 4457–4462.

[16] Y. Zhang, Y. Shen, H. Wang, J. Yong, X. Jiang, On secure wireless communications for IoT under eavesdropper collusion, IEEE Trans. Autom. Sci. Eng. 13 (3) (2016) 1281–1293.

[17] J. Chen, L. Yang, M. Alouini, Physical layer security for cooperative NOMA systems, IEEE Trans. Veh. Technol. 67 (5) (2018) 4645–4649.

[18] D.-T. Do, M.-S. Van Nguyen, Impact of untrusted relay on physical layer security in non-orthogonal multiple access networks, Wirel. Pers. Commun. 106 (3) (2019) 1353–1372.

Security and privacy in wireless body sensor networks using lightweight cryptography scheme

A. Sivasangari[a], A. Ananthi[a], D. Deepa[a], G. Rajesh[b], and X. Mercilin Raajini[c]

[a]*School of Computing, Sathyabama Institute of Science and Technology, Chennai, India,*
[b]*Department of Information Technology, MIT Campus, Anna University, Chennai, India,*
[c]*Department of ECE, Prince Shri Venkateshwara Padmavathy Engineering College, Chennai, India*

Chapter outline

1 Introduction

Wireless body area networks (WBAN) have gained substantial research interest recently, owing to the advent of wireless technologies. The WBAN serves the society by tracking the health status of the patients irrespective of time and location, by means of biosensors. Biosensors are the special kind of sensors, which are power constrained. They are wearable or implantable into the human body. They can transmit data and perform simple manipulations with the collected medical data. The data transmission of WBAN consists of level 1, level 2, and level 3 communication. The secure data transfer among the sensor nodes is performed in level 1. In level 2 the communication between the sensor head (SH) and base station (BS) is performed. The communication between the BS and the remote users is performed in level 3.

43

Security and privacy issues in IoT devices and sensor networks. https://doi.org/10.1016/B978-0-12-821255-4.00003-1

The main issue to be considered during data transmission is the security of the data, which can be transmitted through wireless medium to the remote server. The attacker can perform different types of security attacks on medical information due to broadcast nature of wireless communication. Additionally, the collected medical data are stored in the remote medical server. Thus an effective security mechanism to safeguard the data against data alteration and deletion is the need of the hour. This sort of data tamper would result in wrong diagnosis by the physician, which is a life-threatening issue. By integrating cryptography and access control mechanisms to provide data privacy, this issue could be resolved. However, establishing security and privacy procedures in WBAN is a challenging task due to its power or energy constraints. Therefore WBAN requires a lightweight security system capable of dealing with such security threats. However, establishing security and privacy procedures in WBAN is a challenging task due to its power or energy constraints. Therefore WBAN requires a lightweight security system capable of dealing with such security threats.

WBAN is a human-specific network that possesses several sensor nodes. It is possible to classify the sensor nodes involved in a WBAN as implanted or worn sensor nodes. These nodes'transmission range is 2 m. Security in WBAN is an important issue to be resolved. Health data are confidential and has to be protected from the adversary to ensure the adversary's inability to tamper or delete the health data. The security solutions proposed for wireless sensor networks are not suitable for WBAN. WBAN requires some specific solutions because of its constraints. The restrictions of WBAN are listed in the succeeding text. A sensor utilizes its power to perform any task such as signal detection and communication. The sensors in WBAN are attached to the human body. The temperature of the sensors shoots up during charge. The rise of temperature value is not advisable as it may irritate the patient. The memory capacity of biosensors is limited. Thus any solution to the WBAN must be lightweight and consume small memory.

The biosensors have limited computational ability, as they have limited memory. Biosensors cannot perform complex computations. Communication in WBAN is based on wireless medium. This may cause various security threats and attacks to WBAN. The hacker eavesdrop the identity of a sensor node, which can act as a trust worthy node. This can complicate the security and privacy of WBAN. WBAN consist of lightweight sensors that have only limited communication and computation resources. The security methods applied in WSN are not suitable for WBAN. Security in WBAN is a major unsolved problem while storing in WBAN or during their data transmission. The wireless channel can be easily subject to many types of attacks. The main goal of the current research is to ensure security by introducing a lightweight cryptographic technique for securing data transmission in WBAN. Authentication is provided to safeguard the confidential medical data. To minimize the energy consumption, it is important to minimize the overhead of data transmission.

The integrity and confidentiality of medical data must be protected from hackers. Unauthorized access of health data leads wrong diagnosis and treatment.

The personal server can be a personal computer or a smart phone. The personal server forwards the medical data to the medical server through the Internet. At the other end, the medical server has the health data and several authorized physicians for treating the patients. The medical server has to enforce strict access control policy. The intended physician alone can gain access to the health information of the patient.

The death rate of humankind is escalating every year, owing to several life-threatening diseases. The death rate can be minimized when these diseases are figured out at an early stage. Early detection of diseases is not possible in all cases. However, continuous tracking of a person's health data can be the solution to the aforementioned issue. The most feasible solution to continuous tracking problem is WBAN, which employs wearable or implantable sensors for monitoring the health parameters of a human. WBAN is one of the best remedial solutions and enhances the quality of human life. Thus it is an inexpensive mean to improve the quality of life.

WBAN is accepted as the best solution for remote healthcare monitoring applications. WBAN is a boon to the society, which can save several lives. This type of network allows the physician to provide proper treatment to the patient, irrespective of the time and location of the patient. The health data of the patients are stored in a medical database and enable maintenance of historical health records. This medical database serves as the primary component through which disease diagnosis is made possible Hence the data in the medical database must be protected by all means, as it is the underlying foundation for further process. The physician would end up in the wrong diagnosis or prescription if the provided medical data are incorrect or altered in between. Mostly the health data of the patients are preferred to be stored in a distributive manner, to escape from the effects of single point of failure. Therefore enforcement of security mechanism the entire system is an essential requirement. Data tamper is a very serious issue in WBAN, as the health data reflects on the treatment and diagnosis of the patient. Data tamper may lead to the wrong diagnosis, which is a serious threat. Therefore an effective security mechanism is needed for WBAN. The data that are to be transmitted must be in an unintelligible format, which can be achieved by encryption.

The most important differences between WBAN and WSN are coverage and the data rate. The wireless communication range of WBAN lies from 1 to 2 m and the data rate is restricted below 1 Mbps. The nodes in WBAN are associated with the human body. They can be up to 20, but the number of nodes in WSN can exceed the value of 1000. In WBAN the keys are derived from physiological signals of the human body. The generated key values can exhibit randomness properties of key generation. The nodes in WBAN are severely constrained with respect to communication ability, memory, and computational overhead. Recharging or replacement of batteries is not possible due to the implantable nature of the nodes. The routing attacks are not concentrated in WBAN due to its communication range. WBAN deals with the health records of patients, which are private and confidential, and so a strict security mechanism must be enforced in the WBAN application. On the contrary the security of other wireless networks depends on the nature of the application. WBAN nodes track the physiological signals of the patient at regular

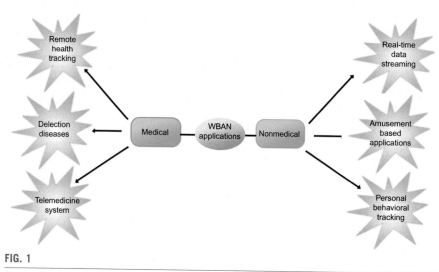

FIG. 1

Applications of WBAN.

time intervals, whereas WSN follows the strategy of event-based tracking. The data transmission in WBAN must be as fast as possible to ensure freshness in medical data. The faster transmission could be achieved at the cost of more energy consumption. Fig. 1 shows the applications of WBAN.

2 Motivation and objective of research

Security in WBAN is an important issue to be resolved. Health data are confidential and have to be protected from the adversary to ensure the adversary's inability to tamper or delete the health data. The security solutions proposed for wireless sensor networks are not suitable for WBAN. WBAN requires some specific solutions because of its constraints. The restrictions of WBAN are listed in the succeeding text.

A sensor utilizes its power to perform any task such as signal detection and communication. The sensors in WBAN are attached to the human body. The temperature of the sensors shoots up during charge. The rise of temperature value is not advisable as it may irritate the patient. The memory capacity of biosensors is limited. Thus any solution to the WBAN must be lightweight and consume small memory. The biosensors have limited computational ability, as they have limited memory. Biosensors cannot perform complex computations. Communication in WBAN is based on wireless medium. This may cause various security threats and attacks to WBAN. The hacker eavesdrop the identity of a sensor node, which can act as a trust worthy node. This can complicate the security and privacy of WBAN.

The previous security issues are to be addressed while designing a healthcare application. There are many challenges to be overcome while designing the mechanism for security and privacy of WBAN. Security and privacy are two essential components in WBAN. Security is achieved by exploiting the security mechanism to provide the security either medical data stored or during their transmission in WBAN. Realizing the importance of security in WBAN, this current research work intends addressing the security issues in WBAN. The present study considers the different problems of security. The main goal of the current research is to ensure security by introducing a lightweight cryptographic technique for securing data transmission in WBAN. Authentication is provided to safeguard the confidential medical data:

(1) to analyze various lightweight cryptography algorithms for designing a new algorithm with minimum computational overhead,
(2) to design a security scheme that derives key values from the biosensors placed in human body,
(3) to design a security solution that satisfies the security requirements of WBAN,
(4) to design a security frame work that can protect data from the unauthorized users.

In this chapter, Section 1 discussed the Introduction of wireless body area network, Section 2 is all about the Reference paper and its issues and security measures, and Section 3 focuses on the proposed model in body sensor network lightweight algorithm and SPLCG algorithm. Section 4 is all about the performance measures and related protocols used in the WSBN, and finally the conclusion, overall summary, and, at last part, the reference papers are included.

3 Related work

Daojing et al. [1] propose a secure, confidential, and DOS-resistant data discovery and data dissemination protocol for WBAN. The symmetric keys are derived from a one-way hash function. The multiple one-way key hash chains can be used for providing authentication. Karur et al. [2] has proposed a wavelet based steganography for protecting the medical information. The proposed technique combines the encryption and scrambling technique that protects the data from attackers. The patient medical information is hidden within the ECG signal.

Ye Yan et al. [3] have proposed novel in-network Advanced Encryption Standard (AES) equivalent mechanism to protect the security and privacy of medical information. The proposed AES equivalent scheme was used to reduce the energy consumption and computational complexity of WBAN. It exploited the use of coordination and assigned the encryption computation workload to the corresponding coordination nodes. The heterogenic residual energy distribution can be used to choose the appropriate coordinator node. The simulation result demonstrates the achievement of the energy efficiency and minimum computational overhead by the proposed scheme. Daojing He [4] has proposed a hash chain–based key updating mechanism and proxy-protected signature technique to provide the secure data transmission and fine grained access control mechanism for accessing the medical information. The network server does the role of original signer. The users play the role of proxy signers. The authorized users can only be able to generate the commands by using proxy signature keys. The updated key values alone are stored in sensor nodes. Both patients and remote users can define the access policy. The keyed hash function is used for performing the data integrity. Alteration of the message is not possible when the value is positive. Otherwise the server discards the message. The proposed system can achieve the requirements of WBAN.

Xiong and Qin [5] have proposed scalable certificate less remote authentication protocol for wireless body area networks. The remote authentication with revocation is achieved by using encryption and signature scheme. The trusted third party is responsible for managing the key values. It is designed to achieve the nonreputation and revocability for communication in WBAN. Zhitao Guan et al. [6] have proposed new secure access control mechanism for cloud-integrated body sensor networks-Mask Certificate Attribute-Based Encryption MC-ABE). A certain signature is intended to hide the plaintext and safely outsource hidden information to cloud severs. A signature and associated privilege authorization certificate is developed, which is used to give information recipient rights. To ensure safety the certificate for each recipient will be masked with a unique value. Thus the certificate is special for every consumer, and the cancelation by removing the mask value can readily be finished. The analysis shows that the scheme proposed meets C-BSN's safety requirements and also has less cost for calculation and storage than for other popular models.

Isma Masood et al. [7] have patient privacy and safety methods suggested in the S-CI. Current methods are categorized in their implementation ranges: multibiometric important generation, pair-specific key establishment, hash function, attribute-based encryption, chaotic maps, hybrid encryption, number theory investigation unit, trimode algorithms, dynamical probability packet marking, and priority-based data forwarding methods. We also offer our six-stage generic framework for privacy and safety of S-CI patient physiological parameters (PPPs), (1) preliminary selection, (2) system entities selection, (3) method selection, (4) PPP access, (5) safety analysis, and (6) performance estimation. Identify and discuss PPPs that are used as datasets and provide the progression of the results of this sector. Finally, conclude with this successful research paper area's open problems and future development. Ahmed Lounis et al. [8] have proposed innovative architecture for big quantity of data generated by medical sensor networks to be recorded and distributed.

Even our own architecture overcomes all the difficulties and makes it simple for medical practitioners to share data in ordinary circumstances and in real emergencies. A highly efficient and highly flexible safety system maintains confidentiality, integrity, and the control of access to automated medical information by fine grain. This mechanism is based on Cipher Text Policy Encryption based on attribute (CP-ABE) for high flexibility and effectiveness. We investigate detailed simulations that demonstrate that in ordinary and emergency situations our system provides effective, slimy, and flexibility access control. Laura Victoria Morales et al. [9] have proposed that this survey studies various proposals that aim to satisfy BAN security requirements, their advances, and remaining challenges. We found that the mentioned requirements have not been comprehensively considered; the majority of the studied proposals do not address the entire BAN architecture; they focus on specific components. Although supporting security of individual BAN components is relevant, a comprehensive security view of an entire BAN environment is needed. Priya Musale et al. [10] have proposed that the focus will be on particular cloud computing issues about healthcare safety and how distinguishing main and key allocation cloud homomorphic encryption helps to meet healthcare regulatory demands. Gaware and Dhonde [11] has giving some idea about fundamentals and challenges, security goals in WSN, and also various security attacks in WSN due to data lost or insecure connection, and they discussed energy consume nodes. Abu-Mahfouz et al. [12] investigated the security in localization algorithm called ALWasHA that enables nodes live malicious reference nodes without disturbing the location estimation.

Bhushan, Bharat, and Sahoo [13] have examined the security threats and vulnerabilities imposed by the distinctive open nature of WSN. In this kind of WSN that they include, both the survivability and security issues and comprehensive survey of various routing and middleware challenges for wireless networks are presented, in different protocol layers. And also various security attacks are identified along with their countermeasures that were investigated, which the researcher has done in this survey. They explained detailed survey of data aggregation and the energy-efficient routing protocols for WSN. Bhushan et al. [14] have proposed the method based on fuzzy-based clustering approach that enables cooperative communication in the network and the balanced load subcluster formation and also Ant colony optimization they used to determine the optimal path to the destination. They did an experimental results and performance analysis to demonstrate that ISFC-BLS is more secure than the heterogeneous ring clustering and other existing clustering techniques, for improving the networks lifetime. Moara-Nkwe, Shi, Myoung Lee, And Hashem Eiza [15] discuss the issues and challenges in the design and implementation of PL-SKG schemes on off-the-shelf WSNs. Then, they proposed a novel key generation scheme to simplify the classic error correcting codes and also diversity of frequency channels available on 802.15.4 compliant nodes to generate keys. Xie, Zheng Yan, Zhen Yao, and Atiquzzaman [16] have provide an overview of WSNs and classify the attacks in WSNs based on protocol stack layers, WSN security measurement, and then research attack detection methods of 11 mainstream attacks. They gather the security data, which is the important role for detecting security anomaly in security measurement.

4 Proposed work

The objective of the proposed research work is described in the succeeding text.

(i) To analyze various lightweight cryptography algorithms for designing a new algorithm with minimum computational overhead.

(ii) To design a security scheme that derives key values from the biosensors placed in human body.
(iii) To design a security solution that satisfies the security requirements of WBAN.
(iv) To design a security frame work that can protect data from the unauthorized users.

Fig. 2 shows the cloud body sensor and the human sensor, which are used to monitor and treat patients remotely and this is considered one of the most promising e-health care techniques. As mentioned in Fig. 3, its work proposes an ECG RC7 algorithm for lightweight safety that makes healthcare apps a high level of safety. The suggested key generation system enables WBAN's adjacent sensor nodes to share the common keys created by the ECG signal. RC7 is a lightweight algorithm used specifically for resource-constrained devices such as tags, encryption, and authentication smart meters. The next area of research is concentrated on lightweight cryptography that has to deal with trade-off between metrics such as safety and security.

4.1 Sensor communication between the sensor nodes

High-pitched QRS complexes represent muscle contraction or ventricular depolarization. Many disturbances, such as noise, electromagnetic interference, and muscle motion, appear in ECG measurement. In the ECG signal, high-pitched QRS complexes represent muscle contraction or ventricular depolarization. Many disturbances, such as noise, electromagnetic interference, and muscle motion, appear in ECG measurement.

The MIT-BIH arhythmia database is downloaded in ECG signals. For key generation the frequency domain analysis is conducted on the ECG signal. The band pass filter is used to remove an ECG signal from the QRS complex. Sampling of the ECG signal for a set 125 Hz and 3 s ECG sampling speed generates 360 samples. The ECG information then carries out a 128-point fast Fourier transform (FFT). The first 64 coefficients are extracted and 12 steps are followed by an exponential quantization feature. The coefficients can be used to generate a 256-bit key value. The SPECG algorithm shown in Fig. 4 is discussed in the following steps:

Step 1:
The ECG Signal feature is extracted and compressed.
Step 2:
This encrypt process will be Secret Key encrypted based on the relevant ECG data and ID.
Step 3:
This step is followed by the DWT time -domain technique for the separation by the sender of the relevant signal and implemented with the compression algorithm Encrypt via the SPECG algorithm.

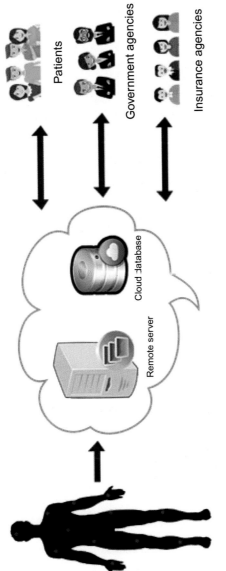

Human

Remote server

Cloud database

Patients

Government agencies

Insurance agencies

FIG. 2

Cloud body sensor.

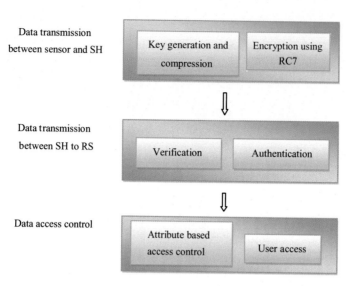

FIG. 3

Phases of proposed model.

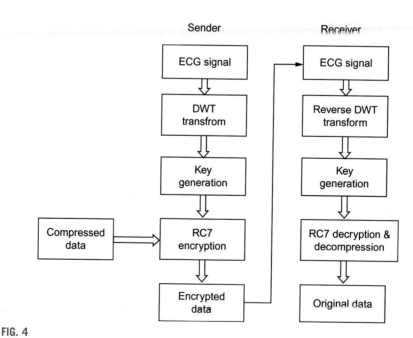

FIG. 4

Proposed SPECG algorithm.

Step 4:

The receiver performs the decryption process in reverse, as if the information were obtained and decrypted using the key and decompressed, and the ECG information were received.

4.2 Encryption and decryption

The art of secret writing and data protection can be defined as cryptography through transformation (encrypting) into a unreadable file, called chip-texts and then transmission through insecure networks, so that nobody except the expected recipient can look into it. Only those with a secret key will decrypt the email to plain text (or decrypt it). Encrypted information, even if contemporary methods of encryption are nearly unbreakable, may sometimes be broken by crypt analysis or code breakage. A number of algorithms contribute to flawless, generally unbreakable encryption outcomes. Huffman compression and SPECG encryption technique are used to transmit the secure ECG signal. RC7 works quicker than RC6 for integers. The Huffman coder based on Lifting wavelet transform, which measure the performance of code with ECG signal compression in terms of signal error and quality. Huffman code provides the high compression ratio, minimum error, and minimum residual difference compare with existing systems. After the compressed data will be encrypted by SPECG encryption mechanism. Compressed data size is normally smaller than the original input signal without loss of data. Finally the ECG signal is transmitted to the receiver side SH without any attack and without any data loss; hence the signal is compressed. The SH forwards the information to personal processing device.

4.3 Secure communication between sensor head to remote server

It is a personal processing device that can receive medical information from a human body sensor unit and transfer it to the cloud server for storage. All the reader device must be registered with medical cloud server. In this phase, mutual authentication between personal processing devices to remote server is performed for preserving anonymity and authentication. Sensor ID information is also stored in the cloud storage. The registration process is described in the succeeding text:

Step 1: PPU will generate the random number N1 that derives the identity of personal processing device. $^{OPU}id = h(^{PPU}id\ N^1\ {}^{T}S||^{K}S)$

Step 2: PPU sends these details to remote server. The server checks if the sequence number is valid or not and simultaneously verifies the ^{PPU}id. Then the server generates a random number S and assigns ^{T}s.

Step 3: The proposed authentication scheme authenticates the PPU by verifying one-time identity and sequence number. The legitimate PPU can form a valid request message M. The server will authenticate the PPU by using $^{h(T}{}_{S},\ ^{K}{}_{S},\ ^{PPU}id,\ ^{N}{}_{1)}.$

In our proposed scheme, one-time identify with the sequence number can give the solution for anonymity and untraceability. It is effective against eavesdropping. The attacker tries to attempt to intercept and modify any legal messages of PPU to pass the verification process to the server. It is very difficult for attackers to know the secret key. The attacker is not able to masquerade the server. It can resist against forgery attack.

4.4 Secure data access in cloud server

To avoid unauthorized access to patient-related information produced by the WBAN, a fine-grained information access policy will be implemented. The source that generated it will not reject a bit of information connected to the patient's source. Access control must guarantee that only authorized entities, users, processes, or devices, will have access to data collected, forwarded and stored by BAN devices. To guarantee access control, two requirements must be addressed. Authentication allows a BAN to establish the identity of a given component, stopping devices that do not belong to a BAN from gaining access to private data. Attackers may pose as a legitimate device, like a sensor or a personal server, to eavesdrop, steal, or send erroneous information, possibly acting sensors and actuators functionality. The algorithm for secure cloud access is described as follows:

In this work, we proposed the fine grained access control scheme for accessing medical data in cloud storage. Medical data can only be accessed if the attributes of user must match with access policy. Every user has his/her own set of attributes. Data user creates the keyword trapdoor for accessing medical data. When he received the ciphertext of health information, the user decrypts the information.

Cloud server provides the required resources to the data users and data owners. This can be managed by cloud service provider. Authentication is performed based on the secret ID issued at the time of registration. This ID is generated by random generator. The service provider encrypts the data before storing in the cloud. Then the user submits an access request to the authentication center to receive the valid token for accessing the medical information. Based on the access policy, valid token along with the decrypted keys are delivered to the authorized user through polynomial scheme.

Step 1: Polynomial coefficients are only shared to the user based on the ID, which is stored in the cloud server.

Step 2: The authentication manager generates the polynomial based on the random key generated by the random generator.

Step 3: The data user also derive the polynomial using coefficients and random key value.

Step 4: If the user is able to construct the polynomial value, only he is the authorized user.

Step 5: Secret key (SK) is shared to the user, and he will decrypt the medical information through the access policy.

5 Performance analysis

FAR and FRR are used for characterizing the proposed scheme. As far as the present study is concerned, the entire work strives to improve the security of the WBAN application. Therefore the standard performance metrics are employed for evaluation of the degree of security. Assuming that many patients are in and around with implanted sensors, the messages passed on by the biosensor should reach the corresponding SH. False Acceptance Rate can be computed by the following formula.

$$FAR = \frac{\text{Wrongly accepted data}}{\text{Total no of attempts}} \qquad (1)$$

FAR is the rate at which the SH accepts messages wrongly from attached biosensors. FAR can be explained by the scenario in which the sensor node of patient P1 tries to communicate with the SH of patient P2 and the SH accepts the data packet. FAR must be the least because it may lead to the wrong diagnosis, which turns out to be a very serious issue.

FRR is the rate at which the SH rejects the message of the biosensors of the same patient. In case a sensor from patient reaches the SH of patient, the SH must reject the request. FRR can be computed by the formula given:

$$FRR = \frac{\text{Wrongly rejected data}}{\text{Total number of attemps}} \qquad (2)$$

At the same time the SH may react in the opposite way by rejecting the data packet being sent by the constituent sensor. This results in adverse conditions, and the FRR must also be the least. HTER is computed by the combination of FAR and FRR. HTER can be calculated by

$$HTER = FAR + FRR/2 \qquad (3)$$

The following Figs. 5–7 show the FAR, FRR, HETR, and encryption time analysis of proposed work.

The current work is evaluated in terms of false acceptance rate (FAR), false rejection rate (FRR), and half error total rate (HETR) of ECG data and tolerance values, which are discussed in Table 1 and plotted in Figs. 5–8; it is inferred that if the tolerance t value increases, the FAR also increases. The possibility of matching two feature sets that does not belong to the same person also increases. In contrast to the FAR, FRR decreases when t increases. The hamming distance can be used to compare the sender and receiver information. Number of bit positions differed in the strings that calculate it. The present research work is compared with existing algorithm.

In the existing IPI scheme, at the sender, features F are extracted from the ECG signals to form a secret k, which is used to encrypt the glucose data or general message, and then send the encrypted message and the HASH-based message authentication code to the receiver. After the receiver gets the packet, it could recover the secret k using the ECG signal measured at the receiver's site and then decrypt the

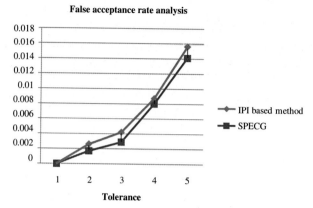

FIG. 5

FAR analysis.

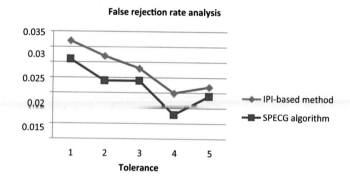

FIG. 6

FRR analysis.

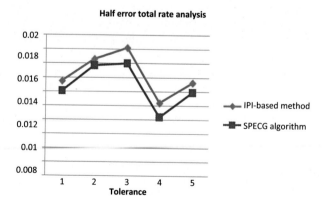

FIG. 7

HETR analysis.

Table 1 Comparison of security parameters of different WBAN protocols.

Security properties	Revocable authentication protocol	Certificate less remote authentication scheme	I-round authentication scheme	Proposed method
Resistant to denial-of-service attack	Yes	Yes	No	Yes
Key compromise impersonation attack	No	No	No	Yes
Privileged attack	No	No	No	Yes
Eavesdropping attack	Yes	Yes	Yes	Yes
Replay attack	Yes	Yes	Yes	Yes
Provides anonymity	Yes	Yes	Yes	Yes
Provides mutual authentication	Yes	No	No	No

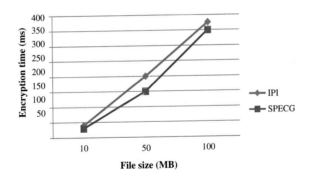

FIG. 8

Encryption time.

encrypted message using the key k. When the message is received at the receiver, the MAC is recalculated from it using the same algorithm. The results will be compared with the received MAC to complete the authentication process.

From the experimental results, it is evident that the performance of the current research work is satisfactory. The encryption time of proposed ECG Hummingbird is 33% faster than Blow Fish algorithm. The suggested algorithm sustains the confidentiality, integrity, and authentication of medical information. The attackers cannot measure the ECG signals. Hence the attackers will not be able to find the key derived from the ECG signal. The security analysis of the present work proved that ECG signal meets the design goals for key agreement, namely, length, randomness, and distinctiveness. If a hacker is trying to send the replay message to the receiver, the receiver will discard it.

6 Summary

The proposed SPECG algorithm ensures secure biohealth science applications based on the ECG signals. As the ECG signals vary from person to person, it is highly motivated for extracting the peak-valued features and to apply intelligent encryption algorithm for secure transmission. Security of the system is achieved through incorporation of an SPECG algorithm. The present study enables secure communication in the network, and the proposed work is lightweight, which is an added advantage to the system. As this system relies on ECG for key generation, the proposed work is more secure. The ECG signal is unique, so, it cannot be duplicated. The proposed algorithm is evaluated in terms of FAR, FRR, and HTER. The system is evaluated with varying level of tolerance. The simulation results show that the proposed work is secured than the existing system. Analysis of the experimental results leads to the conclusion that the proposed work requires the least time for carrying out encryption compared to any other algorithms.

References

[1] D. He, S. Chan, A novel and light weight system to secure wireless medical sensors networks, IEEE J. Biomed. Health Inform. 18 (1) (2014) 316–326.

[2] S. Kaur, O. Farooq, R. Singhal, B.S. Ahuja, Digital watermarking of ECG data for secure wireless communication, in: Proceedings of the International Conference on Telecommunication and Computing Recent Trends in Information, 2010, pp. 140–144.

[3] Y. Yan, T. Shu, Energy-efficient in-network encryption/decryption for wireless body area networks, in: IEEE Symposium on Selected Area on Communications, 2014, pp. 2442–2447.

[4] D. He, C. Chen, S. Chan, B. Jiajun, P. Zhang, Secure and lightweight network admission and transmission protocol for body sensor networks, IEEE J. Biomed. Health Inform. 17 (3) (2013) 664–674.

[5] H. Xiong, Z. Qin, Revocable and scalable certificateless remote authentication protocol with anonymity for wireless body area networks, IEEE Trans. Information Forensics Secur. 10 (7) (2015) 1442–1455.

[6] L.V. Morales, D. Delgado-Ruiz, S.J. Rued, A comprehensive security for body area networks: a survey, Int. J. Netwk. Sec. 21 (2) (2019) 342–354.

[7] Z. Guan, T. Yang, X. Du, Achieving secure and efficient data access control for cloud-integrated body sensor networks, Int. J. Innov. Eng. Technol. 8 (3) (2017) 254 ISSN: 2319-1058.

[8] I. Masood, Y. Wang, A. Daud, N.R. Aljohani, H. Dawood, Towards smart healthcare: patient data privacy and security in sensor-cloud infrastructure, Published 4 November- Wirel. Commun. Mob. Comput. 2018 (2018) 2143897.

[9] G. Rathi, M. Abinaya, M. Deepika, T. Kavyasri, Healthcare data security in cloud computing, Int. J. Innov. Res. Comput. Commun. Eng. 3 (3) (March 2015).

[10] A. Lounis, A. Hadjidj, A. Bouabdallah, Y. Challal, Healing on the cloud: secure cloud architecture for medical wireless sensor networks, Futur. Gener. Comput. Syst. 55 (2016) 266–277 Elsevier.

[11] G. Atul, S.B. Dhonde, A survey on security attacks in wireless sensor networks, in: 2016 3rd International Conference on Computing for Sustainable Global Development (INDIACom), IEEE, 2016, pp. 536–539.

[12] A.M. Abu-Mahfouz, G.P. Hancke, Evaluating ALWadHA for providing secure localisation for wireless sensor networks, in: 2013 Africon, IEEE, 2013, pp. 1–5.

[13] B. Bhushan, G. Sahoo, Recent advances in attacks, technical challenges, vulnerabilities and their countermeasures in wireless sensor networks, Wirel. Pers. Commun. 98 (2) (2018) 2037–2077.

[14] B. Bhushan, G. Sahoo, ISFC-BLS (intelligent and secured fuzzy clustering algorithm using balanced load sub-cluster formation) in WSN environment, Wirel. Pers. Commun. (2019) 1–28.

[15] K. Moara-Nkwe, Q. Shi, G.M. Lee, M.H. Eiza, A novel physical layer secure key generation and refreshment scheme for wireless sensor networks, IEEE Access 6 (2018) 11374–11387.

[16] H. Xie, Z. Yan, Z. Yao, M. Atiquzzaman, Data collection for security measurement in wireless sensor networks: a survey, IEEE Internet Things J. 6 (2) (2018) 2205–2224.

Impact of thermal effects on wireless body area networks and routing strategies

4

B. Banuselvasaraswathy[a] and Vimalathithan Rathinasabapathy[b]

[a]Department of ECE, Sri Krishna College of Technology, Coimbatore, India, [b]Department of ECE, Karpagam College of Engineering, Coimbatore, India

Chapter outline

61

Security and privacy issues in IoT devices and sensor networks. https://doi.org/10.1016/B978-0-12-821255-4.00004-3

1 Introduction

Technological advancement in today's modern world leads to progress of wireless body area networks (WBAN)–based human health monitoring system to change the lifestyle of human beings by providing health care services such as medical data access, memory enhancement, medical monitoring, and communication in emergency conditions. The health condition of patient is monitored continuously by utilizing two types of sensors, namely, implantable and wearable sensors [1]. Hence, continuous monitoring helps to detect emergency conditions and diseases at high-risk patients in prior. The wearable devices should be compact, lightweight, and flexible for normal human body's movements making it comfortable for the patients to wear for a longer duration to increase the accuracy of measured signal [2]. The benefits of wearable devices are composed of improved quality of life, increased patient safety, cost-effectiveness, and increased battery longevity. Additionally, the investigation of remote device includes data monitoring and device interruption.

The processes involved in sensor unit are sensing, data aggregation, computations, and other transceiver operations. Among these operations, transmission and reception of signals consume more amount of energy because of continuous transceiver operation. Moreover, heat is dissipated during energy consumption resulting in damage of nearby heat-sensitive parts of the body. Moreover the congestion problem due to nonregulated traffic and retransmission also increases the temperature of the node due to excessive power consumption [2]. Likewise, variable data rate is also another crucial factor to be considered in routing protocols because high data rate is required for application utilizing multimedia content and less data rate is desired for acquiring data from temperature sensors.

In this work a novel algorithm named OPOTRP is proposed. This algorithm focuses on maintaining the node temperature below the threshold value with minimum critical data delay time between source and destination. Furthermore the quality of service (QoS) for critical data was improved with reduced data delivery delay. Also, maintaining the temperature below threshold value will prolong the node lifetime in WBAN. The operations of the proposed algorithm are as follows: (1) The shortest route and temperature of each node are calculated. If the distance between the source node (SN) and destination node (DN) is higher, then the relay node is selected. (2) Three different priority levels such as low, medium, and critical are assigned to the sensed data. (3) The node temperature is continuously monitored;

if the node temperature exceeds the threshold value, then the node transmits only high priority signals (level 3) from the SN to DN. Other signals are transmitted through an alternative path until the node's temperature reaches to normal temperature.

The remaining content of the paper is organized as follows: In Section 2, several conventional thermal-aware routing protocols were discussed in detail. Section 3 highlights thermal influence on medical WSN. Section 4 describes newly proposed OPOTRP along with its operating conditions. Likewise, Section 5 explains the simulation results obtained for the corresponding proposed algorithm followed by conclusion.

2 Thermal-aware routing protocols

This section discusses the conventional thermal-aware routing protocols introduced by several researchers to analyze the temperature of the sensor nodes and to provide alternative routing path to minimize energy consumption. Heat is dissipated during energy consumption, and increased heat dissipation may cause damage to the tissue of the human body. The change in temperature leads to dynamic performance of sensor that in turn degrades the battery's efficiency and increases data loss at harsh environmental conditions. Hence, it is essential to consider the influence of temperature on sensor nodes.

Thermal-aware routing algorithms are introduced to rectify the aforementioned issues and to promote energy and heat minimization. In thermal routing algorithm the heated nodes are identified and prevented from a part of communication network until it is cooled. Meantime, alternate shortest routing path excluding "hot spot" nodes are chosen to reach the destination, thereby minimizing the amount of heat produced within the network. As mentioned earlier the temperature above the threshold may influence the human tissue and sensor performance. Likewise, it may influence routing protocol and degrade battery's energy leading to increased data loss and affecting the reliability of the network. However, many researches were carried out in routing protocols focusing on delay constraints and power efficiency, but only few researchers considered the effects of heat dissipation from nodes and environmental influence on battery's module. This made the researchers to march toward thermal routing protocols that remain still challenging in WBAN. The conventional thermal routing algorithms (as shown in Fig. 1) proposed by several researchers are as follows:

2.1 Thermal-aware routing algorithm (TARA)

Tang et al. proposed TARA [3] to lessen the feasibility of temperature rise of implanted/wearable biomedical sensor within the human body. In this algorithm the overheated nodes (hot spots) are prevented from the normal routing path during communication. The temperature increase potential (TIP) parameter is calculated for

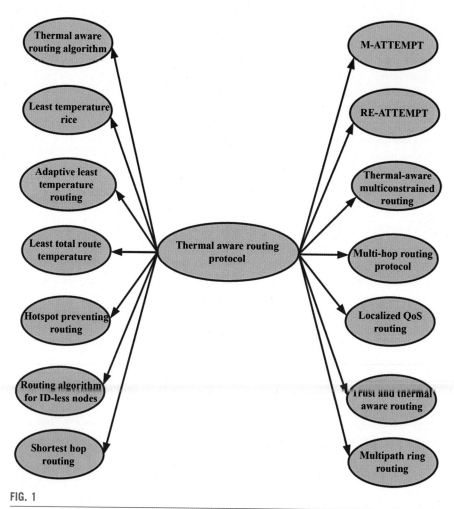

FIG. 1

Thermal-aware routing protocol.

each node on the basis of SAR. Each node notices the temperature of neighboring nodes by determining communication radiation, power consumption, and number of packets. Once a node's temperature reaches beyond the defined threshold value, it does not involve data transmission. TARA ensures that equal temperature is disseminated among all the nodes present in the routing path. As temperature is the only metric considered, TARA faces issues such as reliable transmission, packet drop, and network lifetime.

2.2 Least temperature rise (LTR)

LTR [4] an extended version of TARA is presented to address the overheating issues and increased delay in biomedical sensors. In LTR, next hop is selected by analyzing the temperature of neighboring nodes, and the node with minimum temperature is chosen for the next hop. In this algorithm, MAX HOP limit is defined to avoid bandwidth consumption. A packet exceeding this limit is discarded. Loop is prevented in the network by maintaining a list of recently visited lower-temperature node. Adaptive Least Temperature Routing (ALTR) protocol [5] is similar to LTR with two limits, namely, MAX HOP and MAX HOP ADAPTIVE. In ALTR, when packet exceeds the predefined minimum hop count value, it does not discard as in LTR; instead, it utilizes shortest hop algorithm (SHA) to deliver the packet to the destination with minimum hop count. In this algorithm, proactive delay mechanism is adopted, once it gets a packet from neighboring node with two routing path to the destination with high-temperature sensor nodes. The current node can delay the process by waiting a unit time to send the packet to the lower-temperature neighbor nodes to cool down the temperature.

2.3 Least total route temperature (LTRT)

Least Total Route Temperature (LTRT) [6] is the enhanced approach of LTR and ALTR. In LTRT the data are transferred through the routing path established from source to destination. The routing path is selected on the basis of the temperature instead of considering only the next hop. LTRT selects a least temperature route to forward data instead of considering the next hop alone. The temperature of sensor nodes is converted into graph weights, and shortest routing path are obtained using Dijkstra's algorithm. The temperature of sensor node increases by 1 unit during data transmission and reception and likewise decreases by 1 unit at inactive state during data transmission. In LTRT, packet drop-and-hop count of each packet is minimized; this in turn reduces the heat of sensors. The drawback in LTRT is that every node's temperature within the network should be known. Additionally, temperature acquired for overhead and energy consumption is not measured.

2.4 Hotspot preventing routing (HPR)

Hotspot preventing routing (HPR) [7] is proposed to avoid hot spot formation in the network and to minimize the delay occurring during data transmission. It is composed of two phases, namely, setup phase and routing phase. During setup phase, each node conveys information about its temperature and the shortest route to destination to create routing table. In routing phase, each node maintains a hop count and transmits data through the shortest path by referring the routing table. However, if the destination is neighboring node then data are transmitted directly to the node else next hop is selected on the basis of temperature (less than or equal to threshold value). The packet exceeding the defined threshold value of hop count is discarded. The threshold temperature is selected by taking the average of all neighboring node temperature including the source node. In all the other approaches, the threshold temperature will be predefined.

2.5 RAIN

Routing algorithm for network of homogeneous and ID-less biomedical sensor nodes (RAIN) [8] aims to reduce the average power consumption and temperature of sensor nodes. It consists of setup phase, routing phase, and status update phase. In the setup phase, temporary IDs are generated randomly to sensor nodes for its operational life, and sink node is indicated with ID "zero." All nodes establish communication using their IDs with hello packet. In the routing phase a packet ID with sensor node ID (N), time of data packet (T), and random number (R) is transferred toward the destination. A network loops are prevented by maintaining a hop count for every packet. If packet's hop count exceeds beyond the threshold value, then it is discarded. List of packet ID is maintained to avoid retransmission in the network. The temperature of each node is known by observing its communication activities. In case if the neighbor node is the destination, then the source node will transmit the data directly, and if not the data are forwarded to neighbor node with less temperature. In third phase (update phase), the sink node sends notification of packet ID to all its neighboring nodes to lessen the power consumption.

2.6 Thermal-aware shortest hop routing protocol (TSHR)

Thermal-aware shortest hop routing protocol (TSHR) overcomes the shortcomings of shortest hop routing (SHR) [9]. SHR considers the average of temperature metrics resulting in high end-to-end delay and decreased network lifetime. TSHR is utilized for applications having high priority data for exchange. If a packet is dropped, it is transmitted again, and copy of transmitted packets is maintained. This protocol decreases the overheating without affecting delay, power consumption, and packet delivery. It consists of two stages like setup stage and routing stage. In setup stage, each node has its own routing table, and in routing stage the nodes transmit the data to the destination using the shortest path. The temperature of the nodes is controlled by establishing two threshold limits, namely, static threshold (T_s) and dynamic threshold (T_{Dn}). T_s is given to all nodes, and if the node temperature is exceeding this limit, then it is neglected from routing path until the node's temperature gets cooled down. T_{Dn} is dynamic threshold that is defined based on the temperature of source node and its neighbor.

2.7 M-ATTEMPT

In [10] mobility-supporting adaptive threshold-based thermal-aware energy-efficient multihop protocol (M-ATTEMPT) is presented to employ heterogeneous sensor nodes. Single-hop communication is preferred for critical data, and multihop communication is utilized for normal data transmission. This algorithm provides features to support mobility and energy management. M-ATTEMPT involves four stages: initialization stage, routing stage, scheduling stage, and data transmission stage. In initialization stage, hello packets are scattered to all nodes. In routing stage, nodes with less hop count and minimum temperature are chosen. In scheduling stage, time-division multiple access (TDMA) is scheduled between the root node and sink node for communication. In data transmission phase the root nodes transfer data to sink node in the specified time slot for normal data delivery. Finally the sink node takes some time to combine the received data.

2.8 TMQoS

In [11], TMQoS is demonstrated to perform cross-layer communication and to maintain adequate temperature level. It is a proactive routing protocol that aims to maintain ongoing routing table by collecting information from the neighboring nodes. The information is collected by sending beacon packet. Beacon packets are mainly used to exchange information between the sensor nodes. As soon as the beacon packets are received through MAC receiver module, the routing table module constructs the routing table by determining the values of temperature, delay, hop count, and reliability. The reliability and hop-to-hop delay of neighboring nodes are estimated through reliability and delay module. Temperature module updates the node's temperature in the routing table. Hot spot avoidance mechanism helps to keep the packet far from hot spot region. The beacon packets are sent from routing table to the MAC transmitter module constituted by beacon packet constructor module.

2.9 RE-ATTEMPT

In [12] RE-ATTEMPT routing algorithm, the sink nodes are arranged in descending order by considering data rate. The nodes having higher data rate are deployed near less mobility region, and nodes with low data rate are placed at high mobility area of human body. The protocol is sectioned into five stages such as initialization, routing, phase, scheduling, and data transmission. At the initialization, distance between root node and sink nodes is calculated by generating "hello message," and all nodes will be aware of routing path, distance, and its neighboring nodes. In the routing phase, efficient routing path with less hop count is selected. In scheduling process the communication of sink and root nodes takes places at fixed time slot. In data transmission stage, data transmission occurs only during specified time slot.

2.10 M2E2 multihop routing protocol

In [13], energy efficient M2E2 multihop routing protocol is established. It is divided into four phases like initialization, routing, scheduling, and data transmission. During initialization phase the distance between the root node and sink node is calculated by broadcasting "hello message." In routing phase, data are sent to medical server. After receiving home signal, one of the implanted nodes present on the human will set up a link with home node routing table. In this way data are sent through direct link from sink node to medical server, and energy is not restricted in direct link as critical data are sent over selected link. On the other hand, if home signal is not received, then the communication link is established between routing table of implanted nodes and multihop link. Moreover, this phase also takes care of hot spot avoidance. Scheduling phase and data transmission phase perform similar to RE-ATTEMPT protocol.

2.11 Thermal aware–localized QoS routing protocol

In [14], thermal aware–localized QoS routing protocol (TLQoS) is proposed. The main objective of the routing protocol is to address the QoS for implanted sensors. This protocol aims at maintaining the temperature at the adequate level preventing damaging of tissue. Additionally, it avoids loop formation and unnecessary routes with large hop count to sink node minimizing delay in the network. Among all neighboring nodes in routing tables, suitable nodes for data forwarding are selected using localized approach.

2.12 Trust and thermal-aware routing protocol

In [15] the author in TTRP aims at providing trust to node, neglecting misbehaving nodes, and ensures hot spot–free communication among implanted biomedical sensor nodes. The TTRP consists of three phases: trust estimation phase, route discovery phase, and route maintenance phase. Trust estimation phase evaluates the trustiness of relay nodes by analyzing the packet forward behavior. This phase is further subdivided into direct trust and indirect trust. Direct trust has a single-hop neighbor, and in indirect trust, nodes with direct trust beyond the threshold are considered. The main function of route discovery phase is to identify the trusted node and establish a routing path free from hot spot in the network.

2.13 Self-healing thermal-aware RPL routing protocol

In [16], self-healing thermal-aware RPL routing protocol is presented. In this protocol, efficient routing path is decided by hot spot to transmit data. The best route is selected by considering the node with less temperature and power. Routing can take place in both forward and reverse path with self-healing, that is, ability of the network to self-repair by itself. This protocol is mainly concentrated on IPv6 routing for lossy networks and low power (RPL).

2.14 Multipath ring routing protocol

Multipath ring routing protocol [17] is presented. The basic idea of this protocol is to create multiple paths to avoid delay. The sensor network is organized into different levels like a ring based on hop distance. The sink node is indicated as "ring level 0," and the next levels increase their ring value. The ring level measures the distance between sensor node and destination using hop count. It is composed of multipath construction phase and data transmission phase. In the first case, request packet marked with ring level 0 is forwarded to all the neighboring nodes to set up connection. After receiving the packet the neighboring nodes will update its ring level by increasing its value to 1 and transmits the packet. This phase takes care of organizing the sensor node into ring level, and values are assigned to each level at the end of the phase. In data transmission phase the nodes are separated and arranged based on their ring level value.

From the earlier explained different thermal-aware routing protocols, it is understood that each protocol tries to rectify the effect of temperature on human tissue by providing alternative routing path to improve the communication between the nodes. But still, several problems are associated with sensor node like data transmission losses, packet losses, node failure due to temperature, limited data sensing, battery failure, high transmission delay, high-power consumption, and path failure.

3 Introduction about the thermal influence on medical WSN

The main function of WBAN is to transmit the collected physiological data of patients to central node for further analysis. In WBAN the measuring element (sensors) and transmission (networks) are highly influenced by the temperature. The sensor node mainly gets heated due to longtime sensing, continuous data transmission/reception, and environmental temperature changes resulting in reduced lifetime and increased data loss. To improvise the solution for these problems, furthermore analysis is required in terms of temperature point of view.

3.1 Problems faced by recent scenario

In WBAN, wearable devices monitor the health conditions of patients continuously. It transmits the measured vital physiological signals like glucose level, blood pressure, pulse rate, and heart beat monitoring to the relay node. The data reach the relay node through the routing path established between source and destination. The nodes transmitting the sensed data to the remote location must take new and shortest route each time to the destination. This leads to temperature rise in implanted nodes for a specific time interval during data transmission. As discussed earlier the increased temperature brings harm to the human tissue. Temperature rise of sensor nodes during data transmission is discussed in the following section with an example of WBAN consisting of 10 nodes. The temperature rise of sensor node is analyzed at different scenario and illustrated step by step. Fig. 2 shows the selection of source–destination nodes in WBAN.

Initially the sensor node will sense the signal and assigns priority level (3, critical; 2, medium; and 1, low). The data are transmitted via shortest priority by referring to the routing table. Fig. 3 shows the establishment of shortest connection path between source and destination. A shortest routing path is established from source node (1) to destination node (11) via intermediate nodes (3–4). The sensed signals are transmitted continuously, and routing table of corresponding nodes is updated.

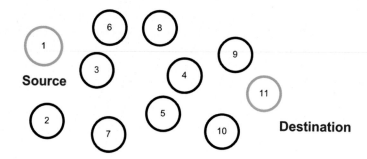

FIG. 2

Selection of source-destination nodes in WBAN.

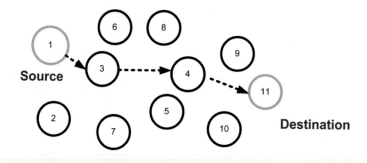

FIG. 3

Establishment of shortest routing path for signal transmission.

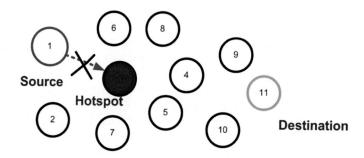

FIG. 4

Hot spot detection in routing path.

Fig. 4 shows the halt of data transmission due to hot spot detection in sensor network; node 3 is termed as hot spot. Therefore an alternative shortest routing path must be selected for further data transmission.

The temperature of the node gets heated due to continuous data flow through the particular node; such nodes are termed as "hot spot." Further data transmission through this node is suspended to avoid damage to the human tissue. In Fig. 5, node 3 becomes "hot spot" due to continuous data flow through it. Hence, other alternative path through either node 2 or node 6 can be selected to prevent damage of human tissue. In Fig. 5, alternative routing path is chosen from 1-6-8-9-11.

As shown in Fig. 6, all the data are transmitted via alternative path chosen to reach destination. Therefore the continuous passage of data through this alternative path (node 6) increases the temperature of node creating hot spot. Fig. 6 clearly highlights node 6 as "hot spot" (red color) due to continuous usage of alternative path.

The node 6 will backtrack the data to source node, and the entire convergence has to be reworked for the entire network. Another alternative path has to be adopted for the rest of the communication. The source node finds an alternative path by referring to the routing table. Once the path is identified, the source node will establish a connection between source and destination node via the alternative shortest path 1-2-7-5-10-11 as shown in Fig. 7.

After sometime, once again "hot spot" is created at node 7 due to continuous data transmission via that particular shortest path. Fig. 8 shows the "hot spot" creation at

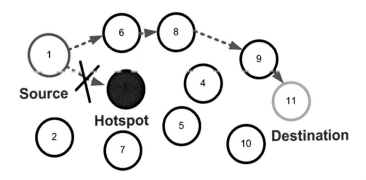

FIG. 5

Data transmission via alternative shortest routing path.

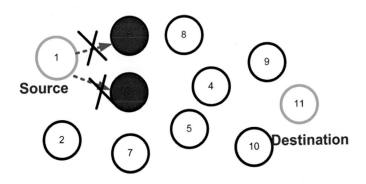

FIG. 6

Detection of hot spot at node 3 due to continuous data transmission.

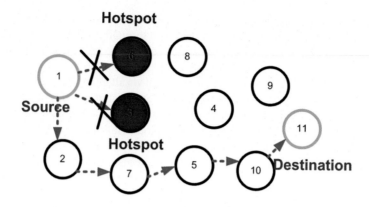

FIG. 7

Establishment of alternative path for signal transmission.

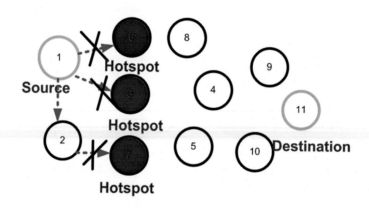

FIG. 8

Detection of hot spot at node 7 due to continuous data transmission.

many alternative path. At this condition the entire communication is halted due to temperature rise. Further, data transmission takes place only when the node's temperature cools down and comes to normal temperature.

3.2 Thermal influence on human tissue

In WBAN, many nodes are placed inside or outside the surface of the human body typically to collect vital signals from human body's critical parts. In general the sensor nodes present within implantable or wearable devices perform monitoring functions. The physiological data collected from these nodes are sent to the sink node or relay node mounted in/on the human body. Thereafter the data are transferred to central point and hub. Further the data are moved to either hospital or professional staff or personal usage or hospital or emergency center. The communication between the sensors scattered in/on the human body is accomplished by wired/wireless techniques. But wireless communications are highly preferred because in wired communication the wires may break and cause inconvenience for patient's mobility.

The wireless communication is highly sensitive to direction, model of antenna, external disturbances, and obstacles [18]. In WBAN the electromagnetic waves are propagated through human body, and it will act as a communication channel for data transmission [19]. Mainly, losses occur because of absorption of power by the tissue that is dissipated in the form of heat [20]. The penetration depth and dielectric properties in the tissue depend on water content that varies based on the type of tissue. Moreover the penetration depth decreases with increase in frequency. If the tissue present in brain, skin, and muscle are high in water content, then the EM waves are attenuated completely owing to high loss and permittivity before it reaches the receiver [18]. Hence, it is necessary to determine the permittivity and conductivity values of the tissues for every individual as it varies depending on anatomy and age [21]. Alternatively, human body's mobility and position adopted by body also play a vital role in received signal strength. It is found from the literature carried out from several researchers that nearly 20 dB of attenuation is due to various relative movements of legs and arms, and line of sight (LoS) also gets affected due to several motion and position of the body [22].

3.3 Wireless communication technologies for data transfer in WBAN

In general, there exist different communication technologies such as Zigbee, ultra-wide band (UWB), Bluetooth, and short-range radio frequency (RF). But all these approaches are not appropriate for data transmission in health-monitoring applications as all these techniques are easily affected by large signal leakage, electromagnetic interference, and eavesdropping [23]. Additionally, RF signal suffers from attenuation and body shadowing effect around human body during data transmission [24]. This leads to data loss and retransmission increasing the power consumption and heat dissipation. Later, human body communication (HBC) also called as intra-body communication was developed. In HBC, human body acts as a transmission medium for data communication. In HBC, achieving low-power consumption is an important factor to be considered during data transmission from wearable and implantable devices to lengthen the lifetime of battery [23]. The sensor nodes are powered by battery. If the battery is drained, then sensor does not remain functional affecting the performance of entire network. However, reduced battery power needs continuous replacement, thereby decreasing the network lifetime. Thus energy conservation is an additional important factor to be considered in battery to prevent data loss.

3.4 Sources of energy consumption

Basically, energy consumption of a node relies on processing operations like sensing, data aggregation, computations, and transceiver operations. Among all operations, transmission and reception operation consume more amount of energy and dissipate heat affecting the sensitive tissues and delicate organs. These thermal effects change based on intensity of electrical signal and on the number of transmission/reception. A transmitter model consists of the following components, namely, amplifier, frequency synthesizer, modulator, and oscillator. Thus the total transmission power is due to power consumption of each component present in the transmitter model [25]. Among all these components, modulator consumes minimum energy, and power amplifier consumes high energy in WBAN nodes. In receiver model the energy consumption is due to oscillator, mixer, frequency synthesizer, demodulator, and amplifier.

3.5 Specific absorption rate (SAR)

The amount of radiation being absorbed by the human body is termed as SAR. It is also defined as the rate of absorption of heat per unit weight. It is given by Eq. (1) [3]:

$$\text{SAR} = \frac{\sigma [F_E]^2}{\rho} \ (\text{W/kg}) \tag{1}$$

where σ is the electrical conductivity, ρ is the density of human tissue or any material, and F_E is electric field induced. The sensors are recharged using radio frequency (RF). The range of RF is 2–20 MHz. Hence the exposure of human body to RF ranges continuously damages the human tissue. The SAR for the human body is given in Eq. (2) [20]:

$$\text{SAR}_{\text{RF}} = \frac{\sigma [F_E]^2_{\text{RF}}}{\rho} \ (\text{W/kg}) \tag{2}$$

Apart from this, antenna radiation and power dissipation also increase the temperature of sensors and deteriorate human tissue. In implanted sensors the total power consumption is calculated based on power dissipation density, sensor's volume, sensor architecture, and enabling technologies [20] associated. The amount of temperature rise is calculated using Eq. (3) [26]:

$$\rho C_p \frac{dT}{dt} = K \nabla^2 T - b(T - T_b) + \rho SAR + P_c + Q_m \ (\text{w/m}^2) \tag{3}$$

where K is thermal conductivity, ρ is the density of mass, T_b is the blood and tissue temperature, b is blood perfusion constant, and C_p is the specific heat of the human tissue. Table 1 gives the general notions used in the modeling.

There are two types of biological effects, namely, short-term and long-term effects. These occur during data transmission via wireless link in the human body [27]:

- Short-term effects—These effects are caused immediately when exposed to RF.
- Long-term effects—These effects are not immediate; prolonged exposure to RF continuously to many years leads to some degenerative diseases.

Table 1 General terms used for theoretical modeling.

Notation	Meaning
Σ	Electrical conductivity
P	Density of material
RF	Radio frequency
SAR	Specific absorption rate
F_E	Induced electric field
dT/dt	Rate of temperature increase
Cp	Tissue-specific heat
b	Blood perfusion constant
K	Thermal conductivity
Qm	Metabolic heating
T	Temperature
Tb	Blood and tissue temperature
Pc	Power dissipation of circuitry
$K\nabla^2 T$	Heat transfer due to the conduction

4 Proposed protocol (OPOTRP)

The lifetime of sensor node in WBAN directly depends on the sensor node temperature. The primary challenge of a network includes quality of service, network performance, stability, efficiency, and optimized routing path with minimum temperature [28]. The data transmission continuously through a particular node causes node failure and increases the network temperature [2, 4, 8]. Similarly the distance from the source to destination is also a main factor for energy consumption and high data loss that influences the QoS and reliability of the network. To overcome these problems, OPOTRP is proposed.

OPOTRP consists of initialization phase, relay node selection, priority assigning, and temperature monitoring.

- *Initialization phase:* The nodes set up initial condition for the network. The connection is established after collecting the information from neighboring nodes such as node temperature and hop count.
- *Relay node selection:* The source transmits the data directly to the destination if the distance is small. At large distance, relay node is selected, and data are forwarded via this relay node. The relay node is selected by considering the following parameters, namely, temperature, battery energy, and distance.
- *Priority assigning:* The input data signals are given priority level such as levels 1, 2, and 3 (low, medium, and critical).
- *Temperature measurement*: All the three level data signals are transmitted once the node's temperature reaches the minimum threshold value. At high threshold value, only level 3 (critical) signals are transmitted. Whereas level 1 and level 2 signals are rerouted to an alternative routing path until the nodes return to normal temperature.

4.1 OPOTRP functional procedure

The proposed algorithm is shown in Fig. 9. It comprises multiple selection and routing procedures to send data efficiently. The proposed system model consists of
sender (source), relay (intermediate) node, and destination in a single routing path.
It helps in managing n-number of routes to send information from sensing node to the
destination node. The proposed model measures the temperature of the relay node

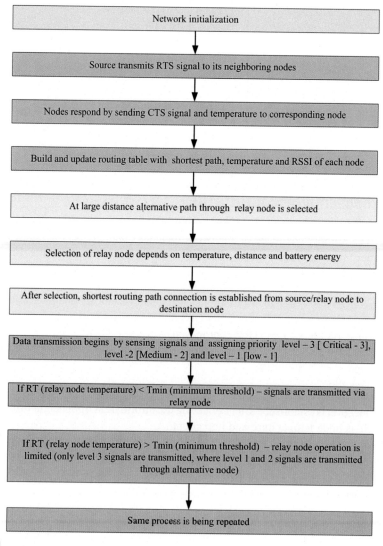

FIG. 9

Functional flow diagram of proposed (OPOTRP).

frequently to ensure that the temperature range lies within the threshold limit of SAR. If the temperature lies within the threshold limit, the data are transmitted through relay node; otherwise an alternative path is chosen to reach the destination. The proposed model consists of several procedures as follows.

4.1.1 Initialization stage

In initialization stage the sensor node send RTS signal to all its neighboring nodes to establish routing path for data transmission. The receiver nodes reply by sending CTS signal and temperature of it to source node to begin data transmission. The temperature of the individual node is calculated to verify the threshold limit of SAR. Furthermore, sensors also collect the status of the battery and traffic of the particular node. After collecting the information from the sensor node, the source establishes shortest routing path to reach destination and commences data transmission.

4.1.2 Relay node selection

At the start of data transmission, all nodes are assumed to have uniform temperature and energy. At small distance the data are directly forwarded to destination. In case of long distance, the paths are routed via relay nodes to the destination. A relay node (RN) is selected by the utility factor. The node with low temperature, maximum energy, and high data rate is chosen as relay node. A relay node should be selected with utmost care because the failure of RN degrades the efficiency and QoS of the network.

4.1.3 Assigning priority level for data in sensor node

The sensed data are classified into three priority levels: level 1 for normal data (low priority), level 2 signals for patients in abnormal state (medium priority), and level 3 signals for patient in critical state (high priority). The other factor to be considered for sending data is RN temperature. It is defined with low threshold and high threshold levels. If the relay node temperature is below the low threshold value, all priority (level 1, level 2, and level 3) data are communicated via selected RN. If the RN temperature reaches above high threshold level, the RN allows only the high priority (level 3) signals until the temperature reaches below low threshold value. However, other priority data (level 1 and level 2) are transmitted through alternative RN to DN.

As the temperature of the sensor node reaches high threshold value, the battery usage of sensor node also increases. This increased battery usage rises the node's temperature resulting in node failure with increased data losses. Thus, if the sensor node reaches maximum threshold value, all functions are restricted for the meantime to reduce the temperature of RN. The normal operation is carried out once the temperature of RN gets cooled down. Thus QoS, network lifetime, and battery performance improve by maintaining node's temperature at nominal range.

4.1.4 Alternative relay node selection

The alternative RN selection also will follow the same procedure as mentioned earlier. The RN will continue their operation without any interruption; at the same time, low priority signals are also transmitted to DN based on RN threshold limits. From this, it is clear that the proposed method's quality of service (QoS) is improved with minimum data loss. If the transmission is done via initial RN, then the transmission time will be minimized with increased lifetime and performance of the network.

4.2 Optimal temperature selection

In WBAN the sensor placed in/on the human body generates heat during several operations such as data transmission, state of battery, and data reception. From several studies, it was reported that SAR is the important factor to be considered. It specifies the RF range at which the human body can be exposed; continuous exposure to the radiation beyond the given range causes damage to human tissue. Thus it is necessary to select an optimal routing path within the sensor nodes to increase the network lifetime and efficiency.

5 Results and discussion

In this section the OPOTRP simulation results are compared with other conventional routing protocols for validation.

5.1 Simulation parameters

The OPOTRP protocol is simulated using MATLAB, and nodes are deployed in area 250×250 m. The total number of nodes does not exceed greater than 150. The data transmission range between source and destination is around 50 m. The room temperature is assumed as 30°C. The minimum threshold temperature is set to 60°C, and the required energy for data transmission is 10,000 µW. The simulation parameters are listed in Table 2. The following assumptions are made before the start of simulation. All the nodes in the simulation network are fixed and uniform with respect to transmission range. The value of node-specific heat and mass is kept at constant value. The heat reduction rate of the node at idle state is set to 2 units.

5.2 Variation in temperature at different times

Fig. 10 depicts the simulation results of temperature variations at different times. The temperature value increases with time interval. The temperature variation of different protocols at different time interval was observed. Initially TARA, LTR, and OPOTRP rise up to 0.2°C, 15°C, and 0.75°C, respectively. At the final stage of the simulation, TARA, LTR, and OPOTRP protocols reach a temperature of 0.6°C, 0.39°C, and 0.3°C, respectively, at the simulation period of 200 s. From the graph,

Table 2 Simulation parameters for OPOTRP.

Parameters	Values
Simulation area	250 × 250 m
Number of nodes	150
Transmission range	50 m
Temperature	30°C
Threshold minimum temperature	60°C
Specific heat of node	0.5 g/J
Required transmission energy	10,000 µW
Hop count	20
Rate of data transmission	250 Kbps
Initial listening time period	1 Unit
Rate of energy transmission	50 nJ/bit
Cooling rate (idle state)	2 units

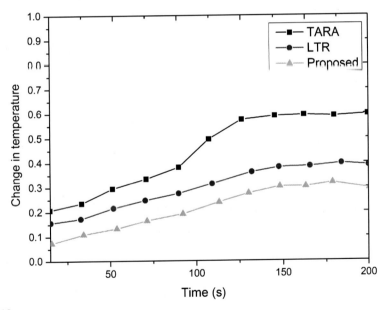

FIG. 10

Temperature versus simulation time.

it is identified that the changes in temperature are negligible once the OPOTRP reaches the threshold value. Likewise, LTR and TARA have higher value as temperature is not considered. Thus maintaining temperature at constant rate can control the temperature rise and data loss.

5.3 Average power consumption

The power consumption can be calculated by finding the difference between initial stage energy and final energy (*t*). The power consumed by proposed OPOTRP, LTR, and TARA is illustrated in Fig. 11. The results clearly show that the proposed protocol consumes less energy compared with LTR and TARA except at high data packet arrival rate. The average data rate of five is maintained during data transmission. Finally the lifetime of proposed protocol is high when compared with the lifetime of TARA and LTR except for high priority level 3 signals.

5.4 Network lifetime analysis

The lifetime of sensor node utilizing different protocols is shown in Fig. 12. The data transmission at high data rate consumes more power with minimum lifetime. The proposed protocols are assigned with three priority levels. In this protocol the data packets with high priority will reach the destination on time. More energy was utilized during data transmission at a rate of five. Furthermore, high data rate is required for transmitting high priority signals than transmitting other two priority signals. This protocol improves QoS with reduced data loss.

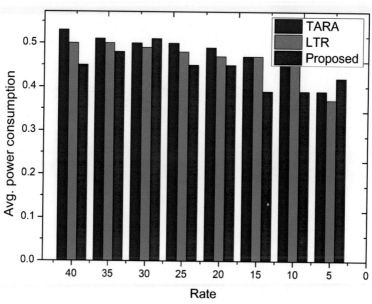

FIG. 11

The average power consumption of different protocol.

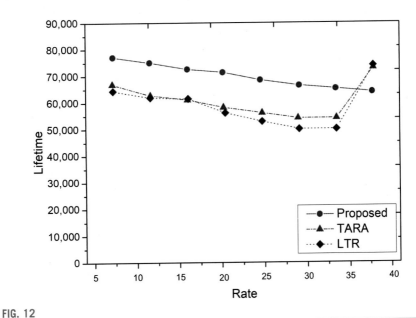

FIG. 12

Lifetime analysis versus data rate.

5.5 Different data priority signal

The amount of energy consumed for packet transmission is less but remains high for transmitting data at high data rate. It is shown in Fig. 11. The proposed protocol transmits critical signals at larger number than other existing protocols (Fig. 13). Therefore it consumes only less energy, and heating ratio is controlled by selecting another routing path [29]. The temperature influences of the average packet receiving ratio of TARA and LTR algorithm that results in increased packet losses at higher rate, minimizing the QoS of the protocol.

5.6 Heating ratio

The amount of heat dissipated should be kept low to avoid tissue damage [30]. This is achieved in proposed method by utilizing power control strategy. In this technique the temperature of sensor nodes is maintained within the threshold limits and is monitored continuously [31]. If the node's temperature reaches maximum threshold, then the signal transmission through a particular node has been limited. A signal only with high priority is allowed to pass through that specific node. The remaining priority signals (levels 1 and 2) are transmitted via alternative node. Moreover the time taken for the sensor node to come back to normal temperature is also less. The same is shown in Fig. 14. The proposed protocol has less heating ratio when compared with other conventional protocols.

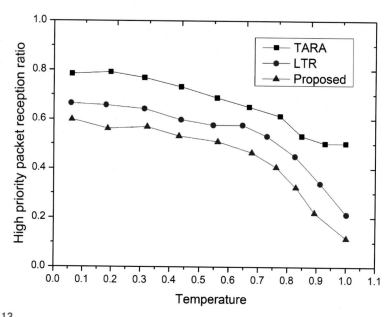

FIG. 13

High priority packet reception ratio versus temperature.

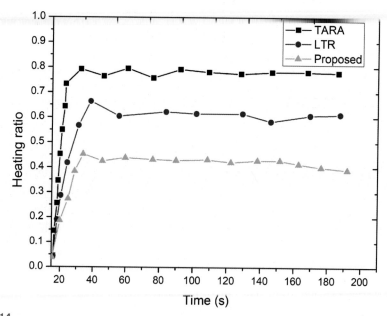

FIG. 14

Heating ratio of sensor node at different protocols.

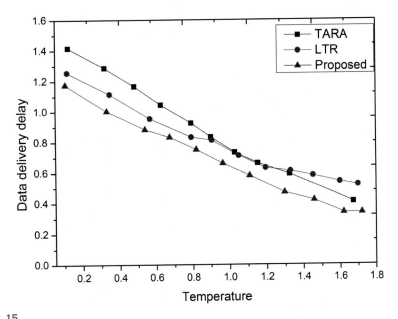

FIG. 15

Comparison of data delivery delay at various protocols.

5.7 Data delivery delay

The amount of packets reach on time is analyzed [32], and same is shown in Fig. 15. In TARA the data packets are not transmitted as the temperature rises. In case of LTR model, as the temperature of node increases, an alternative shortest routing path is identified. However, as the node temperature reaches maximum threshold value, the amount of data forwarded is limited. But high priority signals (level 3) are transmitted without any delay in proposed OPOTRP.

6 Conclusion

The evolution of WBAN has prevented the sudden death of patients by diagnosing and treating the diseases earlier. This is accomplished by the implantable/wearable devices placed inside/on the surface of the human skin. One of the major challenges faced by the WBAN is the temperature rise of sensor node. Thus OPOTRP was introduced to minimize the heat dissipation. The proposed algorithm minimizes the sensor node temperature by utilizing predefined threshold limits. The distance between source to destination is calculated, and the nodes with shortest routing path are preferred. For large distance, relay node is chosen as an intermediate node to minimize delay. Besides, priority levels are assigned to the sensed signals, and the data with highest priority levels are transmitted without any delay. The OPOTRP is compared with existing protocols such as TARA and LTR. From the simulation results, it is clear that the OPOTRP is efficient in terms of delay, heat dissipation, and temperature with improved network lifetime.

References

[1] G. Schirner, D. Erdogmus, K. Chowdhury, T. Padir, The future of human-in-the-loop cyber-physical systems, Computer 1 (2013) 36–45.

[2] A.K. Yetisen, J.L. Martinez-Hurtado, B. Ünal, A. Khademhosseini, H. Butt, Wearables in medicine, Adv. Mater. 30 (33) (2018) 1706910.

[3] Q. Tang, N. Tummala, S.K. Gupta, L. Schwiebert, TARA: thermal-aware routing algorithm for implanted sensor networks, in: International Conference on Distributed Computing in Sensor Systems, Springer, Berlin, Heidelberg, 2005, pp. 206–217.

[4] A. Bag, M.A. Bassiouni, Energy efficient thermal aware routing algorithms for embedded biomedical sensor networks, in: 2006 IEEE International Conference on Mobile Ad Hoc and Sensor Systems, IEEE, 2006, pp. 604–609.

[5] C. Oey, S. Moh, A survey on temperature-aware routing protocols in wireless body sensor networks, Sensors 13 (8) (2013) 9860–9877.

[6] D. Takahashi, Y. Xiao, F. Hu, LTRT: least total-route temperature routing for embedded biomedical sensor networks, in: IEEE GLOBECOM 2007-IEEE Global Telecommunications Conference, IEEE, 2007, pp. 641–645.

[7] A. Bag, M.A. Bassiouni, Hotspot preventing routing algorithm for delay-sensitive applications of in vivo biomedical sensor networks, Inform. Fusion 9 (3) (2008) 389–398.

[8] A. Bag, M.A. Bassiouni, Routing algorithm for network of homogeneous and id-less biomedical sensor nodes (RAIN), in: 2008 IEEE Sensors Applications Symposium, IEEE, 2008, pp. 68–73.

[9] F. Ahourai, M. Tabandeh, M. Jahed, S. Moradi, A thermal-aware shortest hop routing algorithm for in vivo biomedical sensor networks, in: 2009 Sixth International Conference on Information Technology: New Generations, IEEE, 2009, pp. 1612–1613.

[10] N. Javaid, Z. Abbas, M.S. Fareed, Z.A. Khan, N. Alrajeh, M-ATTEMPT: a new energy-efficient routing protocol for wireless body area sensor networks, Procedia Comput. Sci. 19 (2013) 224–231.

[11] M.M. Monowar, M. Mehedi Hassan, F. Bajaber, M.A. Hamid, A. Alamri, Thermal-aware multiconstrained intrabody QoS routing for wireless body area networks, Int. J. Distrib. Sens. Netw. 10 (3) (2014) 676312.

[12] A. Ahmad, N. Javaid, U. Qasim, M. Ishfaq, Z.A. Khan, T.A. Alghamdi, RE-ATTEMPT: a new energy-efficient routing protocol for wireless body area sensor networks, Int. J. Distrib. Sens. Netw. 10 (4) (2014) 464010.

[13] O. Rafatkhah, M.Z. Lighvan, M2E2: a novel multi-hop routing protocol for wireless body sensor networks, Int. J. Comput. Netw. Commun. Secur. 2 (8) (2014) 260–267.

[14] M. Monowar, F. Bajaber, On designing thermal-aware localized QoS routing protocol for in-vivo sensor nodes in wireless body area networks, Sensors 15 (6) (2015) 14016–14044.

[15] A.R. Bhangwar, P. Kumar, A. Ahmed, M.I. Channa, Trust and thermal aware routing protocol (TTRP) for wireless body area networks, Wirel. Pers. Commun. 97 (1) (2017) 349–364.

[16] C.M. Amrita, D. Babiyola, Self healing thermal aware RPL for body area networks, Int. J. Sci. Environ. Technol. 2 (2017) 1143–1152.

[17] G.M. Huang, W.J. Tao, P.S. Liu, S.Y. Liu, Multipath ring routing in wireless sensor networks, in: Applied Mechanics and Materials, 347 Trans Tech Publications, 2013, pp. 701–705.

[18] M. Vallejo, J. Recas, P. del Valle, J. Ayala, Accurate human tissue characterization for energy-efficient wireless on-body communications, Sensors 13 (6) (2013) 7546–7569.

[19] B. Kibret, A.K. Teshome, D.T. Lai, Human body as antenna and its effect on human body communications, Prog. Electromagn. Res. 148 (2014) 193–207.

[20] Q. Tang, N. Tummala, S.K. Gupta, L. Schwiebert, Communication scheduling to minimize thermal effects of implanted biosensor networks in homogeneous tissue, IEEE Trans. Biomed. Eng. 52 (7) (2005) 1285–1294.

[21] A. Peyman, C. Gabriel, E.H. Grant, G. Vermeeren, L. Martens, Variation of the dielectric properties of tissues with age: the effect on the values of SAR in children when exposed to walkie–talkie devices, Phys. Med. Biol. 54 (2) (2008) 227.

[22] M. Di Renzo, R.M. Buehrer, J. Torres, Pulse shape distortion and ranging accuracy in UWB-based body area networks for full-body motion capture and gait analysis, in: IEEE GLOBECOM 2007-IEEE Global Telecommunications Conference, IEEE, 2007, pp. 3775–3780.

[23] J.F. Zhao, X.M. Chen, B.D. Liang, Q.X. Chen, A review on human body communication: signal propagation model, communication performance, and experimental issues, Wirel. Commun. Mob. Comput. 2017 (2017).

[24] A. Fort, J. Ryckaert, C. Desset, P. De Doncker, P. Wambacq, L. Van Biesen, Ultra-wideband channel model for communication around the human body, IEEE J. Sel. Areas Commun. 24 (4) (2006) 927–933.

[25] S. Cho, Energy Efficient RF Communication Systems for Wireless Microsensors, Doctoral DissertationMassachusetts Institute of Technology, 2002.

[26] H.H. Pennes, Analysis of tissue and arterial blood temperatures in the resting human forearm, J. Appl. Physiol. 1 (2) (1948) 93–122.

[27] V. De Santis, M. Feliziani, F. Maradei, Safety assessment of UWB radio systems for body area network by the method, IEEE Trans. Magn. 46 (8) (2010) 3245–3248.

[28] F. Jamil, M. Iqbal, R. Amin, D. Kim, Adaptive thermal-aware routing protocol for wireless body area network, Electronics 8 (1) (2019) 47.

[29] N. Bradai, E. Charfi, L.C. Fourati, L. Kamoun, Priority consideration in inter-WBAN data scheduling and aggregation for monitoring systems, Trans. Emerg. Telecommun. Technol. 27 (4) (2016) 589–600.

[30] G. Ahmed, D. Mahmood, S. Islam, Thermal and energy aware routing in wireless body area networks, Int. J. Distrib. Sens. Netw. 15 (6) (2019) p. 1550147719854974.

[31] L.Z. Maymand, V. Ayatollahitafti, A. Gandomi, Traffic control thermal-aware routing in body area networks, J. Soft Comput. Decis. Support Syst. 4 (4) (2017) 17–22.

[32] T. Hayajneh, G. Almashaqbeh, S. Ullah, A.V. Vasilakos, A survey of wireless technologies coexistence in WBAN: analysis and open research issues, Wirel. Netw 20 (8) (2014) 2165–2199.

Four-way binary tree-based data gathering model for WSN

5

Gaurav Bathla[a], Rohit Kumar[b], and Rajneesh Randhawa[c]

[a]Computer Science & Engineering, Chandigarh University, Mohali, Punjab, India [b]Computer Science, Govt. P.G. College, Naraingarh, Haryana, India [c]Department of Computer Science, Punjabi University, Patiala, Punjab, India

Chapter outline

1 Introduction

Wireless sensor network (WSN) is a system of sensor nodes (SNs). These SNs have the capacity to transmit the assembled information from the observed field through wireless connection toward resource opulence base station (BS). The BS can be placed either inside the field of sensing area or at a far distance from the sensor field [1] as governed by design of the application. The BS and SNs can be static or movable as per the application type.

In WSN, clustering is perhaps the best strategy for gathering the SNs alongside the group head (CH). These SNs work at the same time in seclusion with other groups. Network can be easily scaled up with the help of clustering because it along

87

Security and privacy issues in IoT devices and sensor networks. https://doi.org/10.1016/B978-0-12-821255-4.00005-5

with the reduction of the size of routing table can also grow the network with different isolated clusters. So it also helps to improve the system lifetime [2, 3].

Information gathered by SNs is normally passed by means of multihop transmission to the sink SN. This operation is termed as converge cast [4, 5], which utilizes many-to-one communication, where an average ad hoc system utilizes a many-to-many communication model. Because of this, WSN and ad hoc systems work on different routing models. Consequently the tree structure is opted by many routing conventions because of its inborn qualities [6, 7]. Majority of applications in WSN are based on the model with one sender and multiple receiver nodes.

Routing techniques based on the tree structure formation (among SNs) in the network is preferred in many techniques because of not maintaining the routing tables at SN level. Each SN simply sense, aggregate, and send the information to their respective parent [8]. In this way, there is no prerequisite of course discovery component. The most suitable structure for picking TDMA scheduling for routing is tree-based structure [6].

System lifetime can also be improved by opting the shortest path routing among sensor nodes where each node is connected via shortest path to its parent that can be a sink or BS. These structures prove to be most energy-efficient and hence have better lifetime [6–8].

Another scheme as was proposed in [9] uses name allocation via depth first search as labeling to every SNs. Plenty of research has been done in balancing the load and network energy factors for better lifetime [10] and utilizing explicit measurements (remaining vitality, link quality, and nearness to the sink) to form the routing path [11], which thus helps to improve the system lifetime. The effectiveness of tree development calculations is also estimated by the runtime and the quantity of packets exchanges among the sensors [12].

This article proposes scalable four-way binary deployment technique for sensor nodes, which enhances the lifetime of the network and operates in respect to height of the tree.

Rest of paper is organized in which next section covers work done in relevant field so far. Section 3 covers network model and routing technique followed by result of the proposed technique w.r.t. existing ones. Last section concludes the paper.

2 Literature survey

Protocol proposed in [13] scatters load among the sensors of the system by connecting them in a cluster. Notwithstanding, changing the job of CH after each round, data are processed and aggregated at level of each SN, which thus transmitted to BS for further processing. It accomplishes multiple times better lifetime over existing methods. Another protocol [14] depended on vitality variation characterized by different thresholds defined based on application. The SNs continue in detecting the surroundings ceaselessly and it transmits when conditions meet the defined criteria.

PEGASIS [15] concentrated on improving the lifetime of the system by creating an information transmission chain among sensor SNs with least transmission cost, framing a flat routing convention. PEDAP [16] is based on MST tree-based chain formation among sensor nodes, which is governed by resourceful BS. After a specific interim, BS recomputes the routing data alongside disposing of dead SNs. Protocol in [17] divides the network into different virtual regions based on their distance from the BS. This protocol uses multihop communication model among SNs of network, and then a single node transmits network aggregated data to BS. MSMTP [18] extends the work presented in [17] by further creating MST among sensor nodes that are virtually divided into three tiers of the network.

Author in [19] focuses on improving PDR of the network along with mobility of SNs and that of sink. Another BS governed protocol was proposed in [20], which focuses to reduce the transmission distance among SNs. It considers the remaining energy of the SNs into consideration for appointing it as CH. Another routing mechanism based on [18] was proposed in [21] to transmit the data via relay nodes to the sink. These relay nodes are of higher energy and were used for the long-distance transmission to BS. Another MST-based scheme was proposed in [22] for heterogeneous sensor systems with mobile sink, which was moving around the network and the spans of data to reach to the destination in five hops only.

A binary tree connectivity-based scheme named BTDGS was proposed in [23], which creates the virtual binary tree connectivity among communicating nodes. The versatile sink moves along a predefined trajectory alongside communicating its data. In [24], author proposed tree-based routing scheme, which was based on grouping the similar nodes in a group. Author focuses on timely delivery of data along with balancing node among nodes of the network. E^2R^2 [25] was another hierarchical cluster-based routing protocol where CH was elected by CH panel. Packets can be transmitted either in single or multihops to the sink. Author in [26] focused on improving PDR and end-to-end delay works by eliminating the routing tables.

SHORT [27] protocol chooses the SN with the most residual remaining vitality as the head, which is responsible for transmitting the aggregated data to the sink. ELCH [28] being a hierarchical protocol elects the CH based on the responses in favor collected from rest SNs of the network. Protocol defined in [29] (EECFP) depends on the rotational approach among SNs to become CH. A SN with the highest residual vitality is chosen as a CH and is pivoted after each round to keep up the balance among SNs. ECHERP [30] chooses group heads not only by current vitality but also by likewise taking thought of future leftover vitality of the SNs. Author in [31] proposed and compared various node connectivity structures with the lifetime of the routing protocols. After keeping the deployment fixed, lifetime of protocol was compared with w.r.t. different routing schemes.

Network generates threats, which can be handled with the help of secure protocols. Protocols defined in [32, 33] secured theft management and can be easily achieved, which in turn can decrease the rate of cybercrime.

3 Network model

In this section, radio model, network characteristics, routing protocol, and flow charts are discussed.

3.1 Energy dissipation radio model

Radio model becomes a standard for WSN and widely referred in various researches [31, 34, 35] and is shown in Fig. 1. This radio model consists of electronic energy (E_{elec}) for running circuitry of the node and transmission power. It is proportional to the distance between sender and receiver as well as packet size. The receiving power is based on running circuitry and packet size. Free space model is employed for short distances, while the multipath fading model is considered for long distance transmissions. Detailed structure is described in Fig. 1.

$$E_{Tx}(k, d) = k*E_{elec} + k*E_{amp}(d) \tag{1}$$

where E_{Tx} denotes energy spent on k-bit packet over the distance d. E_{amp} is amplification energy required for the message and is given by

$$Eamp(d) = \begin{cases} E_{amp_{fs}} . d^2 \text{ if } d \le d0 \\ E_{amp_{mp}} . d^4 \text{ if } d > d0 \end{cases} \tag{2}$$

d_0 is threshold distance and is calculated as.

$$d_0 = \sqrt{Eamp\_fs/Eamp\_mp} \tag{3}$$

where receiving circuitry consumption depends on message size only and is computed as follows:

$$E_{Rx}(k) = k*E_{elec} \tag{4}$$

$$E_{Da}(k) = 5*10^{-9}*k \tag{5}$$

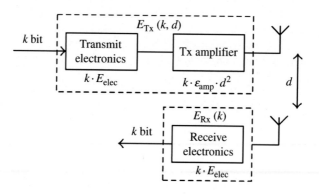

FIG. 1

Energy radio model [31].

Symbols used in radio model are as follows:

Symbols used

Symbol	Meaning
E_{elec}	Electronic energy for running circuitry
E_{Tx}	Transmission cost
$E_{(amp\_fs)}$	Amplification energy free space
$E_{(amp\_mp)}$	Amplification energy free space
d	Distance between communicating nodes
k	Packet size in bits
E_{Rx}	Receiving cost of a packet
E_{Da}	Data aggregation energy

3.2 Network characteristics

i. Nodes are deployed uniformly in the square-shaped area.

ii. Nodes are static, and every node knows its own and its parent location.

iii. The network is virtually divided into four areas by cutting the square region diagonally as shown in Fig. 2A and B.

At each level of the tree, all nodes are assumed to work parallel and isolated from other nodes.

3.3 Proposed (virtual 4-way full binary tree) structure

Proposed technique adopts binary tree structure due to following facts:

i. Binary trees are considered as one of the most efficient data structure for organizing and accessing the data. The proposed technique incorporates the arrangement of nodes in the form of the full binary tree that is formed from four sides. In this structure, BS is assumed to be in the center of the region, which is acting as a common root node for the trees from all four sides.

ii. This structure will enhance the lifetime of the network, and it will take very limited time to send the data to BS. Moreover, by extending the network to one more level, it will lead to increase the data transmission hop count by one unit only, and additional 2^{n+1} nodes will be added to the network.

iii. This approach is best suited for the applications where sensor nodes are deployed manually with human intervention, for example, installing sensor nodes in a chamber before starting a nuclear chain reaction.

According to the area of each node and structure of the full binary tree, sensor nodes are divided into the following categories:

i. Root node: Node in the center of the field act as BS.

ii. Parent node: Node at the topmost position in the hierarchy

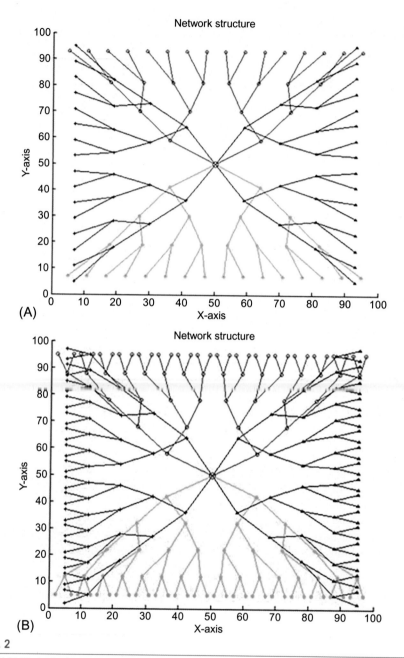

FIG. 2

(A) 121 Node deployment structure in 100*100 area. (B) 249 Node deployment structure in 100*100 area.

iii. Same generation: Nodes that are at the same level as that of the tree (equal hop count from BS).

iv. Leaf Nodes: Nodes at last level of the structure with no descendants.

v. Same area nodes: Virtual area of the network represented in form of the tree.

All sensor nodes are considered to be static, acquainted with location and area id at the time of their deployment. Therefore each node possesses information about its concerned communicating nodes (sender and the receiver from them). However, to classify them as per the aforementioned categories, the following calculations has been used:

i. Root node: Node with Id $= 1$ act as a parent from trees of four sides.

ii. Parent node: All nodes of the network with id $= \lfloor node\_id/2 \rfloor$, where node_id is any node from the network.

iii. Same generation: Children of a parent are said to be siblings, and all nodes at the same height of tree are said to be of the same generation.

iv. Leaf nodes: Nodes for which id*2 is not available in the tree.

v. Same area nodes: Nodes that are assigned same area id.

3.4 Setup of the network

The nodes are uniformly distributed over the square-shaped region. Node distribution starts from the root node (located in the center of the region) toward outer side in each virtual region. Therefore the nodes are approximately equidistant from each other and uniformly dispersed in the field of interest.

The proposed structure is illustrated in Fig. 2 where every full binary tree is depicted by a different color.

Less density of the nodes in the center and greater toward the extreme corners is used because the center part of the network contains BS, and when we move away from the reporting place, more nodes are deployed to monitor the area effectively.

There is also some overlapping between two different trees toward the corner sides, which is retained to maintain generality for deploying the nodes in the region of interest.

$$\text{Number of nodes in the tree} = 2^{h+2} \tag{6}$$

where h is the height of the tree.

Before starting the exchange of the data, each node should be provided with information about its parent, which will help in receiving the transmitted data from the node. By virtue of this setup, a connection will be established between the sender and receiver that will make the proposal a connection-oriented protocol.

3.5 Working of the proposed scheme

Fig. 3A and B shows two different operational stages of the proposed scheme where 3(a) represents working of the network when nodes at level 3 are operational and the rest are assumed to be in sleeping mode. Fig. 3B represents the next stage when nodes

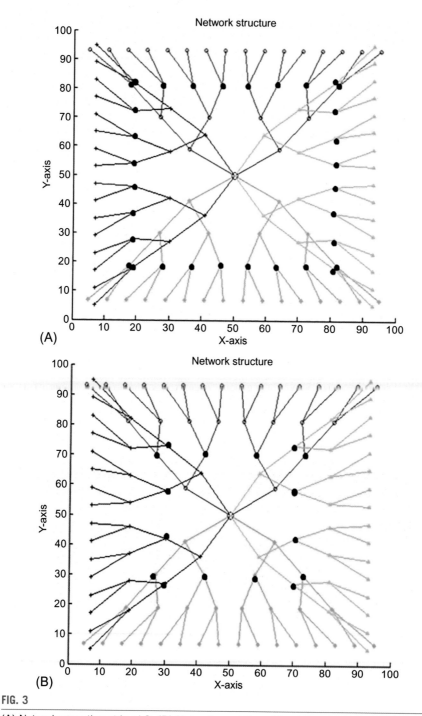

FIG. 3

(A) Network operation at level 3. (B) Network operation at level 2.

at level 2 are operational. These nodes are shown by highlighting the active nodes at that moment.

The sensed data will flow in the direction from leaf node to BS (center of the graph) by each level of incrementing hop count toward the parent node in the following steps.

i. All nodes sense the network data.
ii. Children nodes transmitting the data to its parent, which aggregate the received and its sensed data to its parent node.
iii. Step ii should be repeated until data reach to resource opulence BS, from which further action based on received data can be taken out.
iv. During each round of operation, record of energy depletion, number of packet dropped, and node dead is maintained.

Fig. 4 depicts branch of the tree that illustrates the path of data transmission from leaf to BS in multihops. This transmission takes place parallel with other sensor nodes of the field along with its aggregating data at each parent node.

So broadly classifying the proposed technique is divided into three different phases:

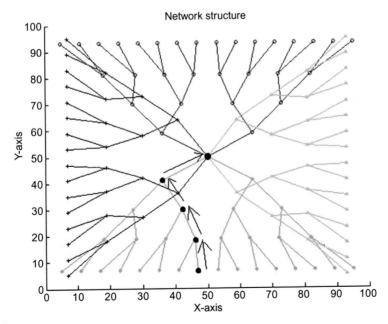

FIG. 4

Transmission path from leaf to BS via intermediate nodes.

3.5.1 Initial phase

The duration of the network setup, in which nodes are arranging themselves in the form a full binary tree. This phase of the protocol follows the routines defined in the next section for setting up the network. It starts the process by **network flow (total and area)**, a process where "total" numbers of nodes are to be deployed in "area." Each node will be following full binary tree hierarchy and will transmit data to its parent node only.

3.5.2 Active phase

This phase comes into action when network operation is started, that is, moment from which nodes start sensing the data and remains into the picture until the last node is in working state. This phase is further divided into two cycles.

a. *With stable period:* The time duration from the start of the active phase till all nodes is in the working state.
b. *With unstable period:* The moment from which the first node dies till the last node is in the operational state. Few adjustments are also required in this period. Whenever a node dies, procedure **TREE-DELETE(T, id)** will discard the node (with identity "id") from the network. This process will maintain the routing structure by connecting the dead node's subtree with the appropriate node.

3.5.3 Dying phase

This phase is considered when results are not assumed to be reliable. Reliability is best when results are obtained from all nodes of the network and are going to deteriorate as each node dies. When 90% of the nodes are dead, then the network is assumed to be in dying state.

3.6 Flow charts

This section defines three procedures (starting with network flow), which give step-by-step working of the proposed technique.

Network Setup (total, area)

Total depicts number of nodes deployed in the field having area of $X*Y$ meter^2. Flowchart depicted in Fig. 5 shows the working of network setup phase.

Network flow (total, area)

Execution of the network functionality in the form of flow chart is illustrated in Fig. 6.

Node Distribution (total, area)

Nodes distribution and initialization of the parameters are depicted in Fig. 7 in the form of flow chart.

TREE-DELETE (T, id)

T is three generated from the previous subroutines and id is the pointer to the node, which is dead. The process of discarding a node is shown in Fig. 8.

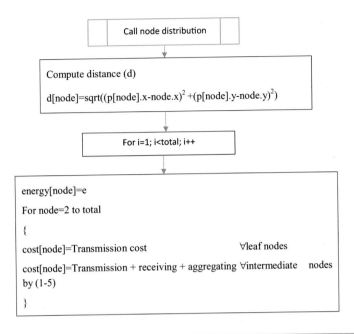

FIG. 5

Flowchart for network setup

4 Result analysis

Proposed four-way binary model is implemented in [36] and is compared with various connectivity structures among sensor nodes. Graphical representation illustrating lifetime of various techniques is shown in Fig. 9 based on parameters defined in Table 1.

As the figure illustrates, VTGDRA outperforms among the compared techniques. But each technique having its own advantages like four-way tree is easily scalable and fast in operation among the compared techniques. So, it depends on the application which routing protocol to adopt for its functioning.

5 Advantages of the proposed scheme

Proposed model expect to achieve the following gaps:

i. This is a static scheme, in which the location of the receiver is predefined and there is no cost for setup phase (in terms of broadcasting messages by CH) of the protocol.

ii. Through this scheme, balance is maintained between transmission time (in terms of hops) and transmission cost (in terms of energy).

iii. The proposed structure is easily scalable (just at the cost of adding one more level in the tree, the network will be able to accommodate total $4*2^h$ (2^{h+2}) nodes where h is the height of the tree).

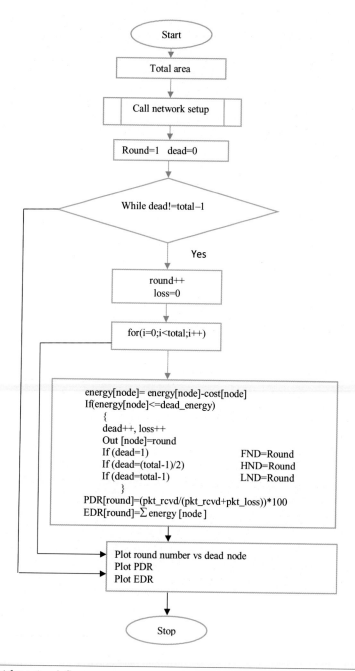

FIG. 6

Flowchart for network flow.

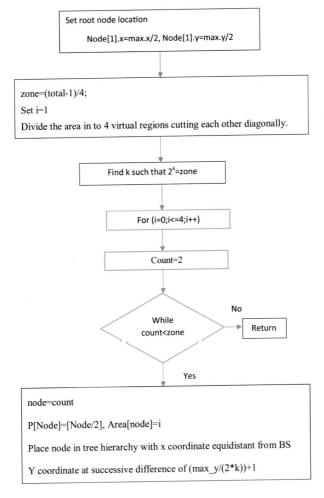

FIG. 7

Flowchart for node distribution.

iv. The proposed model is more reliable because more number of nodes are dedicated to particular area of network. So accuracy of data is more.

v. All sensors are uniformly dispersed in the sensing area with equal number of nodes in all the subtrees. So, maintaining even load distribution on sensors of the network.

vi. Every node has prior information about the receiver node to which it has to transmit the data. So the proposed routing protocol is said to be connection oriented protocol.

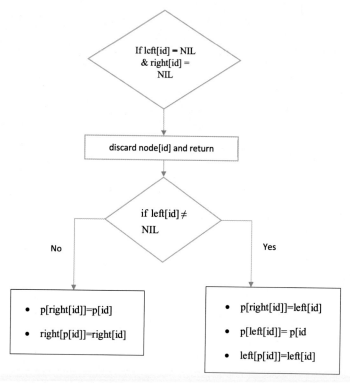

FIG. 8

Flowchart for discarding of node.

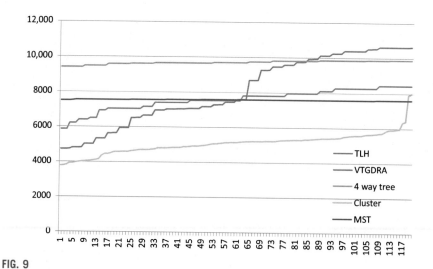

FIG. 9

Lifetime analysis of different structure.

Table 1 Simulation parameters.

Parameter	Value
E_{elec}	50 nj/bit
$E_{Data\ aggregation}$	5 nj/bit/signal
$E_{free\ space}$	10 pj/bit/m2
$E_{multi\ path}$	0.0013 pj/bit/m4
Packet size	2000 bits
Sensing area	100 * 100
BS(X, Y)	50,50
Number of nodes	120
Network energy	192 J

6 Conclusion

The proposed scheme is successful in achieving faster transmission to BS along with good lifetime with respect to the existing schemes. This is due to uniform distribution of all nodes in the network, which are connected in form of full binary tree. It gives an advantage of nodes being well dispersed and properly connected. Every node knows the position of its parent for data forwarding, which results in no overhead of maintaining routing tables. The most important advantage of the proposed scheme is that no node is too frequently used in routing path. This prevents faster draining out, which leads to a better stable lifetime of the network.

Proposed protocol requires parallel operation of all sensor nodes, which are at the same level of the tree. Thus ardent care should be taken so as to avoid data collision. As the proposed scheme follows a tree structure, so any node failure will result in failure of all its descendants.

References

[1] I.F. Akyildiz, W. Su, Y. Sankarasubramaniam, E. Cayirci, A survey on sensor networks, IEEE Commun. Mag. 40 (8) (2002) 102–114.

[2] W.I.S.W. Din, S. Yahya, R. Razali, M. Nasir Taib, A.I.M. Yassin, Energy source saving approach using multi-tier network design technique, in: Control and System Graduate Research Colloquium (ICSGRC), IEEE, 2015, pp. 49–54.

[3] J. Yang, Y. Lin, H. Li, L. Hong, An energy-efficient data gathering algorithm based on clustering for wireless sensor networks, in: Electronics, Communications and Control (ICECC), 2011 International Conference on, IEEE, 2011, pp. 1305–1308.

[4] O.D. Incel, A. Ghosh, B. Krishnamachari, K. Chintalapudi, Fast data collection in tree-based wireless sensor networks, IEEE Trans. Mob. Comput. 11 (1) (2012) 86–99.

[5] T.-S. Chen, H.-W. Tsai, C.-P. Chu, Adjustable convergecast tree protocol for wireless sensor networks, Comput. Commun. 33 (5) (2010) 559–570.

[6] Z. Eskandari, M.H. Yaghmaee, A.H. Mohajerzadeh, Energy efficient spanning tree for data aggregation in wireless sensor networks, in: Computer Communications and Networks, 2008. ICCCN'08. Proceedings of 17th International Conference on, IEEE, 2008, pp. 1–5.

[7] T. Niwat, T. Yoshito, S. Kaoru, Tree-based data dissemination in wireless sensor networks, in: Proceedings of the IEICE General Conference, vol. 2005, Institute of Electronics, Information and Communication Engineers, 2005, p. 42.

[8] N. Mitton, T. Razafindralambo, D. Simplot-Ryl, I. Stojmenovic, Towards a hybrid energy efficient multi-tree-based optimized routing protocol for wireless networks, Sensors 12 (2012) 17295–17319.

[9] E. Chávez, N. Mitton, H. Tejeda, Routing in wireless networks with position trees, in: International Conference on Ad-Hoc Networks and Wireless, Springer, Berlin, Heidelberg, 2007, pp. 32–45.

[10] F. Benbadis, J.-J. Puig, M. Dias de Amorim, C. Chaudet, T. Friedman, D. Simplot-Ryl, Jumps: enhancing hop-count positioning in sensor networks using multiple coordinates, arXiv (2006) preprint cs/0604105.

[11] E.H. Elhafsi, N. Mitton, D. Simplot-Ryl, Cost over progress based energy efficient routing over virtual coordinates in wireless sensor networks, in: World of Wireless, Mobile and Multimedia Networks, 2007. WoWMoM 2007. IEEE International Symposium on a, IEEE, 2007, pp. 1–6.

[12] A. Caruso, S. Chessa, S. De, A. Urpi, GPS free coordinate assignment and routing in wireless sensor networks, in: INFOCOM 2005. 24th Annual Joint Conference of the IEEE Computer and Communications Societies. Proceedings IEEE, vol. 1, IEEE, 2005, pp. 150–160.

[13] W.R. Heinzelman, A. Chandrakasan, H. Balakrishnan, Energy-efficient communication protocol for wireless microsensor networks, in: System Sciences, 2000. Proceedings of the 33rd Annual Hawaii International Conference on, IEEE, 2000, pp. 1–10.

[14] A. Manjeshwar, D.P. Agrawal, TEEN: a routing protocol for enhanced efficiency in wireless sensor networks, in: Proceedings of the 15th Parallel and Distributed Processing Symposium, IEEE Computer Society, San Francisco, 2001, pp. 30189–30195.

[15] S. Lindsey, C.S. Raghavendra, PEGASIS: power-efficient gathering in sensor information systems, in: Aerospace Conference Proceedings, 2002. vol. 3, IEEE, 2002, pp. 1–6.

[16] H.Ö. Tan, I. Körpeoğlu, Power efficient data gathering and aggregation in wireless sensor networks, ACM SIGMOD Rec. 32 (4) (2003) 66–71.

[17] A.S. Ibrahim, Z. Han, K.J. Ray Liu, Distributed energy-efficient cooperative routing in wireless networks, IEEE Trans. Wirel. Commun. 7 (10) (2008) 3930–3941.

[18] G. Khan, G. Bathla, W. Ali, Minimum spanning tree based routing strategy for homogeneous WSN, Int. J. Cloud Comput. Serv. Archit. 1 (2011) 22–29.

[19] N. Hassanzadeh, O. Landsiedel, F. Hermans, O. Rensfelt, T. Voigt, Efficient mobile data collection with mobile collect, in: Distributed Computing in Sensor Systems (DCOSS), 2012 IEEE 8th International Conference on, IEEE, 2012, pp. 25–32.

[20] J.-Y. Chang, P.-H. Ju, An energy-saving routing architecture with a uniform clustering algorithm for wireless body sensor networks, Futur. Gener. Comput. Syst. 35 (2014) 128–140.

[21] G. Bathla, Minimum spanning tree based protocol for heterogeneous wireless sensor networks, i-Manager's J. Wirel. Commun. Netw. 1 (4) (2013) 12–22.

[22] R. Sudarmani, R. Vanithamani, Minimum spanning tree for clustered heterogeneous sensor networks with mobile sink, in: Computational Intelligence and Computing Research (ICCIC), 2015 IEEE International Conference on, IEEE, 2015, pp. 1–6.

[23] C. Zhu, H. Zhang, G. Han, L. Shu, J.J.P.C. Rodrigues, Btdgs: binary-tree based data gathering scheme with mobile sink for wireless multimedia sensor networks, Mobile Netw. Appl. 20 (5) (2015) 604–622.

[24] N. Bagga, S. Sharma, S. Jain, T. Ranjan Sahoo, A cluster-tree based data dissemination routing protocol, Procedia Comput. Sci. 54 (2015) 7–13.

[25] H.K.D. Sarma, A. Kar, R. Mall, Energy efficient and reliable routing for mobile wireless sensor networks, in: Distributed Computing in Sensor Systems Workshops (DCOSSW), 2010 6th IEEE International Conference on, IEEE, 2010, pp. 1–6.

[26] L.K. Wadhwa, R.S. Deshpande, V. Priye, Extended shortcut tree routing for ZigBee based wireless sensor network, Ad Hoc Netw. 37 (2016) 295–300.

[27] Y. Yang, H.-H. Wu, H.-H. Chen, SHORT: shortest hop routing tree for wireless sensor networks, Int. J. Sensor Netw. 2 (5) (2007) 368–374.

[28] J.J. Lotf, M.N. Bonab, S. Khorsandi, A novel cluster-based routing protocol with extending lifetime for wireless sensor networks, in: Wireless and Optical Communications Networks, 2008. WOCN'08. 5th IFIP International Conference on, IEEE, 2008, pp. 1–5.

[29] A. Allirani, M. Suganthi, An energy efficient cluster formation protocol with low latency in wireless sensor networks, World Acad. Sci. Eng. Technol. 51 (2009) 1–7.

[30] S.A. Nikolidakis, D. Kandris, D.D. Vergados, C. Douligeris, Energy efficient routing in wireless sensor networks through balanced clustering, Algorithms 6 (1) (2013) 29–42.

[31] A.E. Tümer, M. Gündüz, Energy-efficient and fast data gathering protocols for indoor wireless sensor networks, Sensors 10 (9) (2010) 8054–8069.

[32] N. Sharma, et al., An intelligent genetic base algorithm for optimal virtual machine migration in cloud computing, Int. J. Recent Technol. Eng. 8 (1) (2019).

[33] L. Pawar, et al., Smart City IOT: smart architectural solution for networking, congestion and heterogeneity, in: IEEE ICICCS, May, 2019.

[34] E.A. Khalil, A. Attea Bara'a, Energy-aware evolutionary routing protocol for dynamic clustering of wireless sensor networks, Swarm Evol. Comput. 1 (4) (2011) 195–203.

[35] J. Wang, J.-U. Kim, L. Shu, Y. Niu, S. Lee, A distance-based energy aware routing algorithm for wireless sensor networks, Sensors 10 (2010) 9493–9511.

[36] G. Bathla, L. Pawar, G. Khan, R. Bajaj, Effect on lifetime of routing protocols by means of different connectivity schemes, Int. J. Sci. Technol. Res. 8 (12) (Dec. 2019) 617–622.

Routing protocols: Key security issues and challenges in IoT, ad hoc, and sensor networks

6

Chanchal Kumar[a] and Shiv Prakash[b]

[a]*Department of Computer Science, Jamia Millia Islamia, New Delhi, India,* [b]*Department of Chemical Engineering, IIT Delhi, New Delhi, India*

Chapter outline

105

Security and privacy issues in IoT devices and sensor networks. https://doi.org/10.1016/B978-0-12-821255-4.00006-7

1 Introduction

The world is rapidly shifting toward a new era of communication where not only human but also every device due to inherent capability to communicate is trying to connect with other devices. "The things or objects communicating with each other through Internet" is termed as Internet of things (IoT). It is an extended version of existing Internet technology connecting each device on the planet. Easy deployment and self-organizing nature without any infrastructure of IoT devices continue to attract the attention of the researcher from academic research and industry. This free flow of information through various communication channels exposes several vulnerabilities of communication models. The basic feature like dynamic topology, wireless links, and operational environment makes IoT devices and sensor networks vulnerable for attacks [1]. The variation in application areas of IOT, along with vulnerabilities of the traditional communication system, introduces new challenges and opens up large scope for potential research. Among various attacks a wormhole is a particularly severe attack that is launched at the time of routing, and it is challenging to detect. Research on the wormhole attack aims to detect it before it harms the network. In this context, this study focuses on the aspects of wormhole-attack prevention. The IoT is a new paradigm and advanced version of existing mobile ad hoc networks (MANETs), which is also self-organized and operates autonomously [2]. The era of MANETs starts from 1972 when the Advanced Research Project Agency (ARPA) sponsored a project recognized as the Packet Radio Network (PRNET). In the early 1980s, this project was further evolved as Survival Adoptive Radio Networks (SURAN). Its primary objective was to provide packet switching to mobile elements in the battlefield, where the infrastructure establishment was not possible [3, 4]. To address security problem and its countermeasures in IoT, mobile ad hoc, and sensor network, this chapter proposed the detection of an algorithm for detection of a severe security attack that is tested in mobile ad hoc network. MANET development can be categorized into three generations. The first generation started in 1972 and known as PRNET.

IoT, mobile ad hoc, and sensor network are emerging as a vibrant area in the networking field. Easy deployment and self-organizing without any infrastructure continue to attract the attention of the researcher from academic research and industry. Host and routing decision functionality by a node in ad hoc network makes uncertainty about the next hop in the communication that makes the system very complex and distributive in nature. The evolution of the smart city and smart devices is making human life easier and more comfortable, but the basic features like dynamic topology, wireless links, and operational environment make networks vulnerable for attacks. The application areas like battlefield, rescue operations, driverless cars, home door security, and police services require security up to full extent, and they always operate in an untrusted environment. The power, security, and routing, the major key issues in the network, require additional care to make network robust. Many researchers have worked in network security; however, classical security solutions are either not compatible or lacks in providing security to smart devices. Through the chapter, we have proposed to detect an acute security attack, that is, the wormhole-attack that is difficult to detect. The goal of the research on the wormhole attack is to detect it before it harms the network.

The major contribution of this chapter is to identify vulnerability of classical transmission time-based mechanism and to eliminate it by developing a novel transmission time-based wormhole detection (TTWD) mechanism. The novel mechanism is developed by enhancing and optimizing existing classical transmission time-based mechanism using dynamic source routing protocol (DSR). Further, a comparative study to evaluate the performance of enhanced mechanism in different scenarios by altering node's speed, node density, tunnel length, and threshold time. When DSR protocol is compared with DSR under wormhole attack by varying speed of nodes, the PDR and throughput on average decreased by 17% and 28%. Further, TTWD was applied that results in growth of PDR and throughput on average by 9% and 15%, respectively. With respect to varying node density, TTWD increased the PDR by 12% and throughput by 14% (on average) as compared with DSR under wormhole attack. With the increment in tunnel length, the detection rate too grew exponentially. When tunnel length exceeds 5, the detection rate approached to approximately 100%. The maximum detection accuracy of TTWD was observed at 35 ms; therefore it was taken as optimal threshold time value.

This chapter covers layer wise architecture for routing, security issues, and challenges. It also includes some existing security threats and models to defend against them. This chapter starts with the introduction and motivation to write this chapter. To understand the problem thoroughly, the critical literature review of routing and security issues, as well as some existing competent solutions, is elaborated in Section 2. Section 3 includes the few counter measures proposed by the author and discussion over the results obtained through simulation of proposed algorithms. Further, critical results analysis with the state of art is discussed in Section 4. Finally the authors will conclude this interesting chapter of a by novel findings and contributions along with future research directions.

2 The security issues, challenges and requirements

In this section various security requirements along with issues and challenges of network security are summarized.

2.1 Network security and requirements

Due to high vulnerabilities toward the security attacks, it is necessary to build a reliable and secure protocol. Any such protocol should satisfy the following requirements listed in the succeeding text [5].

Confidentiality: The data sent by the sender should be understandable to the receiver for which it is being sent. If any intruder node captures the data that are being sent, it should not be able to understand the actual meaning of the data.

Availability: Ensures that networks should be working all the times despite the denial-of-service (DoS) attacks. Whenever an authorized user wants to access the service, it must be available. It should survive even under various attacks.

Integrity: The data sent by the sender should be reached in its original form at the destination, that is, there should not be any tempering with the sent data.

Nonrepudiation: there must be some mechanism to certify that the sender and receiver of the data cannot deny later about data send or receive.

2.2 Security issues and challenges

The networks can be damaged by active attacks and by passive attacks. We have to consider a lot of parameters like environment, radio transmission range, and management security criticalities having different impacts on the security of the network [6].

Operational environment: MANETs are used in a very different environment as compared with a normal network. Some application areas are so critical that they require a foolproof secure network. A silly mistake may result in a disaster. To capture secret information, enemies always attempt to damage or snoop the network. In the operational environment like the battlefield, identification of actual nodes becomes very difficult because nodes always move in and out of the network.

Infrastructure: MANETs are infrastructureless, that is, means no central authority have the power to monitor or direct the data flow over the network. Every node works as host and the router. Every time a new node links the network, it announces its presence and listens from the channel for acquiring the necessary information. Any adversary node can easily come into the network.

Resource: MANETs operate on low bandwidth channels, and battery power is the only source of energy. Because of power constraints, computational power of nodes is kept low. These constraints become big hurdles in the implementation of complex cryptography-based security mechanism.

Topology: Mobility causes topology to change very frequently. Sometimes different ad hoc networks mix together. Then, it leads to IP address duplication.

Shared broadcast channel: Each node in the transmission range can receive the data broadcasted to other nodes. An adversary node can easily analyze the traffic and capture the relevant information [5, 6].

3 Classification of attacks

Interrupting the normal working of the system, damaging the data sent over the network channel, and monitoring the traffic on the network that is not intended for them are considered as attacks. In MANET based on different criteria's, the attacks are classified into several categories. The classification is not mutually exclusive, that is, some attacks can fall in more than one category.

3.1 Based on the attacker's location

The attacker may be the participant of the networks or maybe an outsider. The attacks may classify into internal attacks and external attacks based on their location in the network [7].

Internal attack: In this attack, attackers are part of the same network, that is, intranet. These attacks are performed with the compromised nodes of the network. Nodes that damage the other nodes are known as compromised or malicious nodes. Involvement of authorized nodes in launching the attack makes them difficult to detect.

External attack: In this attack, attackers are not part of same network, that is, intranet. They are easy to detect as compared with internal attacks. These kinds of attacks can be prevented with the help of firewalls and standard security mechanism.

3.2 Based on tempering with data

Active attacks: An external or internal attack that alters or destroys the data being transferred over the channel is known as an active attack. In this kind of attack, integrity requirements of security are violated. The wormhole attack, black-hole attack, byzantine attack, session hijacking, repudiation, etc. fall under this category [8].

Passive attacks: Detection of such attacks is difficult because the normal operation of the network is not disturbed. In eavesdropping and traffic analysis, adversary node snoops ongoing transmission collects secret information without disturbing the network. The confidentiality requirement of security may be violated, if the adversary decrypts data completely and successfully capture actual information. To cope with such attacks, a powerful encryption mechanism should be adopted [8].

4 Attacks and countermeasures on different layers

There are different types of attack that exist for a particular layer in the network protocol stack. However, some attacks can be launched at any layer known as multilayer attacks.

4.1 Physical-layer attacks

Some existing attacks in this layer are jamming and eavesdropping, which not only is confined to this layer but also may be launched at other layers.

Jamming: This attack is launched primarily by military operation and is less significant to the common world. Adversary node sends a signal to the frequency that is same as at which the victim node is receiving from the packets. First, trespasser snoops ongoing transmission to get the correct operational frequency and then disrupts the communication [9, 10].

Eavesdropping attack: It is the passive attack that monitors the transmitted data over the broadcast channel. The attacker captures the secret information without disturbing its normal operation. Strong encryption techniques must be used to prevent these attacks [10].

4.2 Data-link layer attacks

The attacks mounted at the data link layer are commonly passive in nature. Attacks launched at data link layer serve to mount active attacks at other layers. They are difficult to detect. Traffic analysis, monitoring, and disruption are common attacks that are launched at the data link layer. The information about network topology, routing protocol, and security mechanism defending the networks is collected by analyzing the traffic on this layer [11].

4.3 Networks layer attacks

For routing, MANETs use either on-demand or table-driven protocols. In both cases, routing is very tedious due to mobility. Each node has to perform the functionality of router and the host. Nodes that are more than one hop away from each other use intermediate nodes to communicate between themselves. The routing process includes route establishment, route discovery, route updating, and data forwarding phases. Attacks can be launched at any time in this process. Some attacks are described in the succeeding text [12].

Black-hole attack (Sink-hole attack): Black-hole attacks are also known as sink-hole attacks. Depending on the intuition of attackers, black-hole attack can be used for a variety of purposes. Some attackers divert whole traffic toward itself for eavesdropping, while others hinder the route discovery process to consume more power, which may result in the demolition of the entire network. The black-hole attack has basically two properties. First, it exploits the routing protocol by injecting false path information. The second, it drops the packets without forwarding those [9]. This attack may be launched either in route discovery or in route-updating phase. A compromised node registers the optimal path through itself by showing the shortest path or traffic-free route whenever route request is received. Then, all the data packets travelling through the network are diverted toward a node that discards them completely.

Black-hole attacks can be avoided by using secure on-demand routing protocols like Secure Expected Transmission Count (SETX), Security Aware Ad-hoc Routing Technique (SART), one-way hash-chains, and Merkle hash-tree [9, 12].

Byzantine attack: A malicious intermediate node works in collusion within the network and selectively drops the packets, forms routing loops, and diverts the data over the nonoptimal path. They are very hard to detect [13].

Resource consumption attack: Extensive use of waste scarce resources of network, that is, battery power and bandwidth by a malicious node, may collapse the entire network. Adversary nodes unnecessarily generate route request packets. A node consuming battery power of other node is termed as sleep deprivation attack, and it falls under this category [14].

Wormhole attack: Most austere attacks in MANETs are launched during packet forwarding phase at the network layer. It involves pairs of attackers and a tunnel (wormhole). Both attackers in a pair first try to get involved into the network and form a tunnel between themselves to pass the packet. Then, one node receives the data from its neighbor and passes to other colluding node through the tunnel. This colluding node extracts information from the data packet and replays it in to the network near the destination. Presence of tunnel shows a short hop length because of which entire traffic is diverted through direct erroneous link. Besides being an attack, wormhole tunnel mechanism can be used as power saver by providing the direct links for data flow from one end to other end, but it requires extra hardware. The countermeasures like packet leash protocol, DELPHI protocol, and directional antennas may prevent wormhole attacks in which some requires hardware support while others can do without it [15, 16].

Routing attacks: Routing in MANETs is a complex process and more vulnerable. The motive of attacker is to disturb the routing by rushing attacks, routing table overflow, route cache poisoning, etc.

4.4 Transport layer attacks

In MANETs, the key functions of the transport layer are to set up a reliable end-to-end connection and packet delivery etc. The entire session can be hijacked due to the vulnerabilities of this layer. Such attacks are known as Session Hijacking attack. The authentication of a node is verified only at the start of the session. Once the session has been established, no further authentication takes place. The attacker takes the benefit of this vulnerability and tries to hijack the session. After the establishment of the session, adversary node snoops IP address of a legitimate node. Furthermore the attacker impersonates the target node to hijack the session [15, 17].

4.5 Application layer attacks

Nonrepudiation is an essential requirement of security protocols. Repudiation attacks are specific to the application layer of the network protocol suite. Repudiation attacks are the denial of participation of the node in any kind of communication. Once a node has sent the data, it cannot deny this. Spoofing is used to carry out repudiation attacks. An adversary node copies the IP address of the node and then takes part in the communication by masking its IP [15, 17].

4.6 Multilayer attacks

Some attacks are not specific only to a particular layer that may be launched at any layer of the network protocol stack.

4.7 Denial of service (DoS) attacks

Blocking authorized users to access the services for which they are entitled to access is called DoS attack. In MANETs, there is no central monitoring authority, that is, every node has similar access rights and processing power. This unique characteristic of MANETs makes it more vulnerable to an attack like jamming attacks and sys flooding. If a large number of adversary nodes are distributed throughout the network and are blocking legitimate nodes from accessing the services, then DoS attacks are termed as distributed DoS attacks [15]. The packets are used to establish the end-to-end connection at the transport layer. Adversary node floods a large number of SYS packets. Then the victim node replies by sending ACK packet to respond to these SYS packets. Afterward, adversary node spoofs the return address of SYS-packet, and then the victim nodes wait for ACK from the adversary. The legitimate node registers this half-open connection in its routing table. The routing table gets overflowed by a large number of such half connections, although there is expiry limit for such pending connections after which these entries are flushed out. But new entries are registered so fast that they overflow the routing table. The victim node rejects a new connection request even if it is from the legitimate node [18].

5 Survey of security issues, threats and defense mechanisms in IoT

Human life is passing through a virtual world of communication where every object is trying to communicate with other objects. Although IoT devices and technologies are growing rapidly, still this vision is under the developing phase. The objects with sensing competence and processing power, coupled with Internet technology evolved the concept of IoT. This unprecedented growth and variation in the type of devices cause severe security and interoperability challenges. The various existing classical communication protocols are unsuitable for IoT devices. Various threats like [19] denial of service attack, Sybil attack, jamming attack, black-hole attack, wormhole attack, gray-hole attack, eavesdropping attack, selective forwarding attack, and hello flood attack spread across different layers of communication layer. Maintaining the interoperability between different layers of network in itself is challenging and creates many issues; therefore, it is necessary to explore each layer of communication network extensively and uncovers potential threats, attacks, and vulnerabilities. In [20] the author proposed an intrusion detection system for wormhole detection on the network layer. It is one of the most harmful attacks launched at the time of routing using the signal strength indicator. The major security issues and challenges for IoT were studied in [21]. Further the author categorized these identified issues using a layered approach and tabulated IoT security problems with their counter measures.

The ease of deployment or the flexibility of any system comes with its own cost. The cost may be monetary or the compromises the developers did during implementation. Eventually, these compromises lead to severe vulnerabilities or damages of the overall security of the system. In [22] the author proposed and implemented time and trust aware IoT routing technique to defend against Sybil and rank attack. This protocol uses the trust-based technique to sense and isolate these attacks to optimize the entire network. The author also demonstrated the comparative study of the proposed algorithm with other existing techniques. According to free lunch theorem [23], "No security mechanism can be full-proof." Sometimes the proposed mechanism lacks to identify or defend against the real threat, while sometimes it considers many false attacks as the real one. In [24] the author tried to indicate this falsification of attacks by proposing mechanism using honeypot in MANET. The mechanism confirms the attack using the history of attacks and a wormhole attack tree (WAT).

Despite the positive social or economic changes, IoT generates some key challenges, which required prompt and systematic attention from the research community and industry. The heterogeneity of IoT devices and the large flow of data between such devices increase the complexity of the system. In [25, 26] the author presented the essential security requirement for secured IoT networks and taxonomy of protocols along with security services like the authentication of user and devices, key management, user identity management, and access control. Furthermore a comparative study of some recent state-of-art security protocols in IoT was covered.

Among these attacks the wormhole attack is important and most challenging in nature. Therefore this chapter primarily focuses on wormhole attack detection and prevention.

5.1 Wormhole attack and it's counter measure

A lot of threats in MANETS and their countermeasures have been identified till now, and wormhole attack is one of them and affects the discovery of routes. In this technique, two malicious nodes positioned at some specific location to form a secret tunnel. Through this tunnel, one node bypasses routing packets toward other colluding nodes and portray it as the shortest path to the target. By using the link, malicious node launches various attacks like selective dropping, reply attack, and eavesdropping.

5.2 Classification of wormhole attacks

Wormhole attacks may be hidden and exposed in nature. In the hidden wormhole attacks, legitimate nodes are unaware about the presence of malicious nodes, while in case of exposed wormhole attacks, malicious nodes itself involved in are forwarding packets, but legitimate nodes are actually unaware about it.

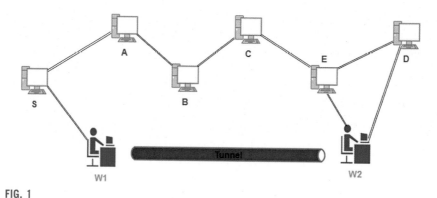

FIG. 1

Wormhole attack [27].

In Fig. 1, S wishes to launch a route to D using the AODV protocol. If RREQ packet is received for the first time, then receiving node updates hop count and puts its identity into the header of the packet.

In the hidden wormhole attacks, malicious nodes do not alter the hop count whereas legitimate node increment the hop-length positively. In the exposed wormhole attack, the attackers update the hop count field. In Fig. 1, both W1 and W2 node updates the hop count and forward. In this scenario, legitimate nodes are attentive to the presence of the wormhole attack, but they do not know which are malicious nodes [16, 27, 28]. The wormhole or tunnel can be out-of-band channel or packet encapsulated channel of the network, which is also known as in-band channel.

5.3 Wormhole detecting and avoiding models

All the previous proposed models either avoid or detect wormhole attacks by modifying the AODV protocol during the route request phase. Otherwise, they require extra hardware and monitoring devices. Some models are described in the succeeding text.

In [29, 30], Packet Leashes Model was proposed to guard against wormhole attack. A leash is a small amount of data that is appended into a packet. It can be geographical or temporal. It is intended to bind the packet's maximum limit to permissible transmission distance. The key purpose behind the packet leash is to limit the transmission range to one hop. Here a node can be authenticated either by specific time-stamp or with location, including a loose time-stamp. It can be easily determined whether a packet has travelled a naive distance. In geographical leash the position of each node is known to other nodes in the connection. While in the temporal leash, special hardware is used to provide tight time synchronization. The temporal leash talks about the upper bound of its lifetime. The temporal leash is implemented through symmetric key cryptography. It needs accurate time synchronization. The main drawback of packet leash mechanism is to obtain extremely tight clock synchronization.

In [16] the author improved the AODV routing technique to countermeasure against the wormhole attack, which is termed as defending against wormhole attack (DAWA). The DAWA uses an artificial immune system and fuzzy logic system to develop a new mechanism. The routes are segregated through the fuzzy logic system. The artificial immune system is used to develop a learning method about wormhole attackers. The proposed scheme is compared with COTA and worm planer algorithms. In [31] an approach based on delay analysis called DELPHI is applied to detect wormhole attack. The DELPHI model can detect both kinds of wormhole attacks. It neither requires any position information, clock synchronization, nor any other special hardware equipment. Therefore it consumes less power. This comprises two different phases; in the first phase source node collects the information about hop-count and disjoint-path delay. Afterwards, in the second phase, the collected information is processed at the source end and compute delay per hop. It is based on the principle that the packet travelled through wormhole experience more delay as compared with the normal path. The DELPHI Request (DREQ) and DELPHI Reply (DREP) packet are used to find the disjoint-path up to the destination. At the source, delay per hop of each route is calculated. The delay per Hop (DPH) is very large in case path contains tunnel as compared with the normal path. Difference between delay per hop under normal condition and under wormhole attacks helps to detect the presence of wormhole. To improve reliability, the process is repeated three times to collect better information. To identify the wormhole all collected DPH values are arranged in descending order. If the difference between DPH_j is greater than DPH_k by threshold value where I and K shows different path, then it shows the existence of the wormhole. DELPHI approach does not work, in case all path results the same delay per hop, that is, all routes are tunneled. This mechanism is only capable of determining the existence of wormhole. The location of wormhole on the path cannot be identified through this approach. In [31] the author proposed an extension of DelPHI, that is, M-DelPHI, to adopt the case of multirate transmission in 802.11 wireless channel. The new protocol M-DelPHI performs and enhances the performance of exiting DelPHI up to 90%.

In [32] the author proposed wormhole avoidance routing protocol (WARP) that uses the concept of multipath routing algorithms. The multiple link disjoint paths are taken into consideration and eventually only one path is used for data transmission. The malicious nodes try to get involved in every disjoint path between source and destination. The WRAP exploits this feature of the malicious node and tried to identify the wormholes attack. Four major modifications are done in basic AODV protocol to implement WRAP. First a new "first hop" field is added in the RREQ frame format of AODV for obtaining disjoint paths. Second the functions of HELLO packet are changed. In AODV protocol, HELLO packets are used to keep the record of neighbors of a node during all the time. Every node continuously receives HELLO packets from its neighbors after a regular interval of time. If the node does not receive HELLO packet from any particular node since a long time, then the entry for that node is flushed out from its routing table. While in WRAP technique if a node receives HELLO packet from any node, it will record the entry of that node in its routing table and accordingly updates the routing table. Third, to collect the information of intermediate nodes, a new concept of decision route reply message (RREP_DEC) is applied, which is sent through the established route. Fourth the format of the routing table is altered to lodge these changes. The anomaly value of every node, that is part of any disjoint path, is calculated on the basis of the information collected through RREP_DEC. Each node keeps a record of neighbor's inconsistent value and if its value exceeds the threshold value, then it indicates that the neighboring node is a malicious one. It has to keep in mind that a legitimate node at the key position in the network has to participate in many paths and, thus, may result in high anomaly value.

In [33] the author proposed a novel protocol to generate Credible Neighbor Discovery (CREDND) by calculating hop difference and local monitoring. To improve efficiency and save power in wireless sensor network, the concept of neighbor ratio threshold (NRH) was proposed. NRH helps to avoid the decision about wormhole detection at each node. The proposed model is compared with two local monitoring and hop difference-based protocols, that is, SEDINE and SECUND.

A mechanism to identify the wormhole attack known as wormhole attack detection protocol using hound packet (WHOP) is proposed by the author in [34]. It doesn't require any hardware support like a directional antenna. There is no need for any précised synchronized clock. It is so strong that it can detect wormhole attack and the malicious nodes forming wormhole. It is based on AODV protocol. Like other protocols, WHOP does not use the nodes that are part of the route. WHOP takes help of remaining nodes in the route-discovery process. The RREP packet format is altered by appending new field to store each node identity on the route. WHOP alters the HELLO packet function from what it used to be in the AODV. The HELLO packet is used to share a node public key among its one-hop neighbor. Once the route has been established, then the source node prepares a hound packet that contains entries of all nodes that are members of the route. The source node computes message digest (MD) of hound packet and signs MD with its own private key. The hound packet is broadcasted into the network. Each node checks its IP address in the hound packet. If a node finds its IP address in hound packet, then it simply forwards the hound packet, otherwise it will process the hound packet. The malicious nodes try to involve in the forwarding process again. The hound packet arrives at the destination node and that the node will process this hound packet. After the processing of the hound packet, the destination node announces whether the route is safe or under wormhole attack.

In [33, 35] author proposed EDWA, where "minimum hop count" is calculated for each node up to the destination node. The end-to-end detection of wormhole attack technique is so strong that it can identify the existence of a wormhole attack on the path and the end points of the wormhole. After the identification of the wormhole end points, information is broadcasted to inform other nodes regarding these malicious nodes. The EDWA is compatible with both the AODV and DSR. It assumes that each node is capable of identifying the geographical location either by using Global Positioning System (GPS) or Global Navigation Satellite System (GNSS) [33, 35].

6 Proposed solutions and analysis

For any efficient reliable and robust communication system like IoT, WNS, and MANET, it is challenging to administer the network layer. Besides other functionalities of the network layer, obtaining secure routing is of utmost priority. In layered architect, various attacks and vulnerabilities like wormhole attack, denial of service, black hole, jamming attack, Sybil attack, eavesdropping attack, selective forwarding attack, and hello flood attack are launched at the network layer during the routing phase. As most of these threats are launched in the routing phase of communication, therefore, the network layer of layer architecture is of primary concern. The wormhole attack is one of the most severe threats launched in route discovery in WSN, IoT, or ad hoc network. In the detection of wormhole attack using transmission time, the detection process completely depends on the correctness of the round-trip time (RTT). The source computes RTT of the packets, that is, route request (RREQ) and route reply (RREP) between the neighbors, while the RTT of every node is calculated by the node itself. The intermediate nodes send back RTT to the source node. The malicious node may add false RTT information in RREP, which may lead to an incorrect computation of RRT between the neighbors. Therefore the malicious nodes are able to hide themselves during the detection process. Therefore a more secure and efficient mechanism is required to preclude the malicious nodes so that they are not able to insert false information in the RREP. To modify the format of the RREP packet and cache of the neighbor node appropriately to store the information required for detection of a wormhole attack. Apply the proposed scheme on the dynamic source routing (DSR) protocol with wormhole attack.

Therefore a model is proposed to detect and prevent the wormhole attack to secure these networks. The testing of the proposed model in the real-world communication networks may affect the entire functionality; therefore it has been tested through simulations using the simulator. This section covers the proposed model in detail and its performance evaluation. To judge the applicability and acceptability of the proposed model, a comparative analysis with other existing models to defend against such threats is also covered.

6.1 Transmission time-based wormhole detection

In [36], transmission time-based mechanism (TTM) is proposed to identify wormhole attack by using ad hoc on-demand distance vector routing (AODV) technique. It is based on the round-trip time (RTT) of the packet. The RTT is computed by using RREQ packet broadcast time and RREP packet arrival time. In transmission time-based wormhole detection (TTWD), TTM is optimized using dynamic source routing protocol with some modifications.

A wormhole attack detection model based on time stamps is known as transmission time-based mechanism (TTM) uses packet RTT. It detects wormhole attack launched on the AODV routing protocol. It is based on the concept that the RTT between malicious nodes forming wormhole is comparatively high than the

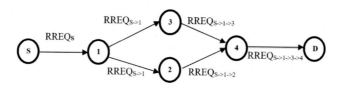

FIG. 2

Route request in TTM.

legitimate nodes. The RTT value between each pair of legitimate nodes is almost the same. Although there is a minor difference between RTT values of the neighbors when one or both are at key position in the network, this difference is much smaller as compared with the threshold value. TTM is tested over the AODV protocol with some modification. The format of RREQ packet is the same as used in the AODV protocol. The source node starts broadcasting a RREQ packet. If RREQ packet is received for the first time, then each intermediate node inserts its node identity in RREQ packet otherwise discards the packet as shown in Fig. 2.

The intermediate nodes further broadcast RREQ packets and keep the record of RREQ sending time in their route cache until RREP packet received [37].

6.1.1 Route reply by the destination

When RREQ packet reached at the destination, the destination node triggers the route reply mechanism. The destination node captures the information about the number of hops RREQ travelled. The simple RREP packet format of the AODV is altered by appending an additional part. When RREP packet is received at intermediate nodes, they calculate RRT between destination and itself. The RTT is inserted in the extensional part, which must be sufficient to store the all RTT values calculated by each node on the path [36, 38].

6.1.2 RTT calculations

Every node keeps recorded time of sending the RREQ packet. On receipt of any RREP packets, surrounding node records its arrival time before further processing. The RTT between source and destination node is evaluated through Eq. (1):

$$(RTT)_{A,d} = |[(T_{RREQ})_A - (T_{RREP})_A]| \tag{1}$$

The RTT between each neighbor node is evaluated through Eq. (2):

$$(RTT_{Nbr})_{A,B} = |[(RTT)_{A,d} - (RRT)_{B,d}]| \tag{2}$$

where

$(RTT_{Nbr})_{A\ B}$: Time between any two neighbors **A** and **B** on the same route
$(T_{RREP})_A$: Time of arrival RREP packet at $(Node)_x$
$(T_{RREQ})_A$: Time of sending RREQ packet from any $(Node)_x$
$(RTT)_{A,\ d}$: Round trip time at any node A and destination
Abs(A - B): Absolute difference between **A** and **B**

Let us consider a path established during the route discovery in any network:

$$S---1---2---W_1---W_2---D$$

Table 1 shows the timing record of broadcasting RREQ packet and receiving RREP packets. Table 1 records RTT for each node is mentioned in the succeeding text.

Each node inserts RTT values in the extensional part of RREP packet and sends to the source. Thus the source gets the RTT time of each node. Now the source node starts wormhole attack detection process and calculates RTT time between each neighbor on the path using Eq. (2) [36, 37].

The value of RTT between the "fake neighbors" is much higher than the value between the real neighbors. The nodes that are not the real neighbor, that is, not in the transmission range of each other, but the presence of wormhole makes them feel that they are neighbors, are considered as "fake neighbor." In Table 2 an average RTT between each pair of nodes is 5, but it is 14 between W_1 and W_2, which shows the aberrant behavior between these nodes. We consider this aberrant behavior as a wormhole attack. Therefore the presence of attack is declared if RTT between two consecutive neighboring nodes is greater than the threshold limit. The performance of the TTM technique depends upon the statistic that how much precisely the calculation of RTT is performed by individual nodes. Therefore, for improved results, experiments are repeated many times although it increases the overhead. Thus there is always trade-off between accuracy and overhead. As the computation of RTT time is performed by the individual node itself and inserted in the RREP packet. Therefore the probabilities of injecting false information in RREP packet by malicious nodes are high. In such a situation, this scheme will not be capable to detect wormhole and consequently the overall performance of the network degrades [36, 37].

Table 1 Sending and arrival records of packet in TTM.

Node's name	RREQ sending time ($T_{(RREQ)}$)	RREP arrival time ($T_{(RREP)}$)	RRT time of node = ($T_{(RREP)}$) - ($T_{(RREQ)}$)
S	0	33	33
1	2.5	30.5	28
2	4	28	24
W_1	7	25	18
W_2	14	18	4

Table 2 RTT_{Nbr} computation at (Node)$_{Source}$.

$(RTT)_{A\ d}$	$(RTT)_{B,\ d}$	$(RTT_{Nbr})_{A\ B}$
33	28	5 (RTT_{S1})
28	24	4 (RTT_{12})
24	18	6 (RTT_{2W1})
18	4	14 (RTT_{W1W2})

6.2 TTWD model

The TTM mechanism showed good performance with respect to bandwidth utilization and can detect the wormhole before it makes any harm to the network. But there is scope to improve RTT mechanism for better accuracy and results. The proposed model is known as (full form) transmission time-based wormhole detection (TTWD) model. The DSR protocol works well even under high mobility of node even though it requires extra memory as compare with other protocols like AODV. Therefore, TTWD model is tested with DSR. Even though some additional features are included in DSR protocol, yet fundamental concepts of route discovery and route reply remains intact.

6.2.1 TTWD algorithm

Main

1. **Begin**
2. Initialization
3. Originator generates the RReq and update the IP address in RRL's originator field.
4. *While(Broadcast(packet P)) do*
     Boolean destination = Receive&ProcessPacket$_{RReq}$(Packet P)
     if(destination)
          Generate$_{RREP()}$
          Break;
     End while
5. *While(Unicast(packet P)) do*
     Boolean Source = Receive&ProcessPacket$_{RRep}$ (Packet P)
     if(Source)
          Break
     End while
6. *Boolen x = DetectWormhole()*
7. *End*

Module 1
Receive&ProcessPacket$_{RReq}$(Packet P)
If $((P_{NodeId} = Self_{nodeId})$ & $(P_{bCastId} = Self_{bCastId})$ & $(P_{HopLen} > = Store_{HopLen}))$ *then*
     Discard Packet
else if$(P_{destNodeId} = Self_{nodeId})$
     return (1)
else
     found = SearchRRLlist(P$_{nodeId}$)
     if (found) *then*
          if (Verify Self$_{nodeId}$ = RRLList[N-1]) *then*
          /* N represents the No. of Node addresses logged in RRL at time T
               Store Req$_{TimeStamp}$ and Discard Packet
          else
               Discard P
          else
               Add Node Entry in RRL

Module 2
>   *Generate$_{RREP}$()*
>   **Reversed RRL and copy it in RREP Packet.**
>   *(To store RTT value, Destination add an extra part into DSR RREP packet)*

Module 3
>   *Unicast&ProcessPacket$_{RRep}$* (Packet P)
>   **If** ($P_{SourceNodeId}$ = Self$_{nodeId}$)***then***
>   return ***True;***
>   *if* (verify Self$_{nodeId}$ = ReverseRRLList [2]) **then**
>   store ***RRep$_{TimeStamp}$*** and forward ***P***
>   *else*
>   *Extract **RRep$_{TimeStamp}$** of **K_{th}** Node for which* Req$_{TimeStamp}$ *was store in **RReq** process*
>   *calculate **RRT** eqn. 3.1*
>   *store **RRT** time at position from where **T_{RREP}** of **K_{th}** node was extracted.*
>   *store **RRep$_{TimeStamp}$** and forward **P***

Module 4
>   *DetectWormhole():*
>   *Extract **RRep$_{TimeStamp}$** of Kth Node for which **Req$_{TimeStamp}$** was store in request process.*
>   *Calculate **RRT** using **Eqn. 3.1***
>   *Calculate **RRT** between each neighbor nodes using **Eqn. 3.2***
>   *For all node pair on route*
>   *If(**RRT**$_i$ > **$T_{ThresholdValue}$**)*
>   *declare detection of wormhole on the route*

6.2.2 Route discovery

In Fig. 3, S$_{Source-node}$ initiates the route discovery by broadcasting a RREQ packet and stores T_{RREQ} *(time of broadcasting Packet$_{RREQ}$)*. The *Packet$_{RREQ}$ contains* sender's address, receiver's address, and broadcast ID.

Nodes 1, 2, and 3 listen to *Packet$_{RREQ}$*. Each mode matches destination address with itself to check whether the packet is sent to it. Their IP addresses do not match with the destination IP address, so all of these nodes will process and inserts its IP address in packet RRL. On further rebroadcasting, the node S snoop back RREQ (see Fig. 4) and recognized that its IP address is stored at $(n-1)^{th}$ position in RRL and, hence, need to store T_{RREQ} of each node, that is, 1, 2, 3, and discard the packet.

Node 4 receives two RREQ one from node 1 and second from node 2. Assume node 4 listen RREQ from node 2 first, then it will drop the RREQ packet from node 1 due to duplicate broadcast ID and originator ID. Likewise, node 5 listens to the *Packet$_{RREQ}$* from nodes 2 and 3. It is assumed that *Packet$_{RREQ}$* from node 2 received first. Both of the nodes match destination (target) IP address again and find neither

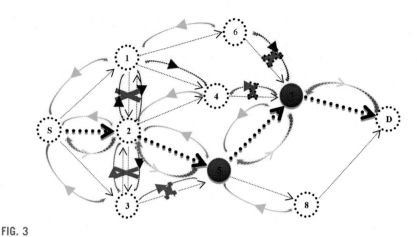

FIG. 3

TTWD route discovery.

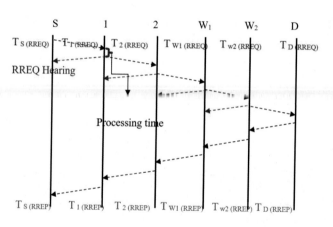

FIG. 4

Timing-diagram of RREQ in TTWD (hearing).

destination nor its address stored at $(n-1)^{th}$ place in RRL; therefore it appends IP addresses in RRL and broadcasts further into the network again.

Similar to source node S, now node 2 will store T_{RREQ} of nodes 4 and 5. A node keeps a record of T_{RREQ} of a particular node till T_{RREP} received at that specific node. Following the same algorithm the RREQ packet will travel to destination node D.

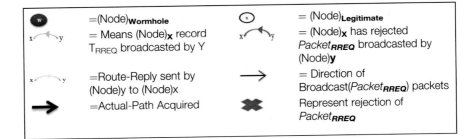

Established route:

$$S \longleftrightarrow 2 \longleftrightarrow 5 \longleftrightarrow 7 \longleftrightarrow D$$

Now (Node)_{Destination} D shall generate RREP. (Node)_{Destination} D alters the format of RREP by adding additional space to store RRT between a pair of nodes. When the RREP packet is unicasted back to node 7, the IP address of node 7 is recorded at second position in reversed RRL (RREP RRL) and therefore stored $(T_{RREP})x$, that is, time of arrival of RREP at (Node)x at the apt index in RREP and forward it. Then, packet is received at next Node 5, which extract $(T_{RREP})_7$ from RREP, that is, time of arrival of RREP at node 7. Node 5 shall calculate the RTT of node 7 using Eq. 1 and insert back in place of $(T_{RREP})_7$. Similarly, other intermediate nodes follow the same steps.

7 Computation of RTT in TTWD

The TTWD enforced to change normal TTM by shifting the calculation of RTT for a particular node from itself to its neighbor. Each node extracts T_{RREP} of its neighbor, calculates RTT and replace back at the place of T_{RREP}. It is presumed that time taken by the packet is negligible to reach to neighbor. In other words, each neighbor will receive a packet at the same time it was broad casted by other nodes (Tables 3–5).

The RTT between each neighbor is calculated the same as calculated in TTM. The round-trip time between nodes 5 and 7 is relatively high and indicate the wormhole attack between them.

Table 3 Time records of RREQ and RREP in TTWD.

Name of nodes	S	2	5	7	D
Name of node who hear RREQ	S	S	2	5	
Time ($TH_{(RREQ)}$) of RREQ hearing	–	2.5	5.5	13	–
$T_{(RREP)}$ time of receiving RREP	30	27	23	15.5	–

Table 4 Node wise RRT computation in TTWD.

Name of nodes	S	2	5	7
Name of node Computing RTT	S	S	2	5
$T_{(REEQ)}$ (RREQ sending time)	0	2.5	5.5	13
$T_{(RREP)}$ (RREP receiving time)	29	26	23	16
RRT of node	29	23.5	17.5	3

Table 5 Computation RTT_{Nbr} in TTWD at source node.

(RTT) A, d	(RTT) B, d	(RTT_{Nbr}) A B
30	24.5	5.5 (RTT_{S2})
23.5	17.5	6 ($RTT_{2\,5}$)
17.5	3	13.5 ($RTT_{5\,7}$)

8 Simulation and experimental evaluation

The performance of the proposed TTWD algorithm in the real-world environment is evaluated through simulation. The simulation is the process to create an illusion that the model is operational in the real-world environment. The simulation environment implemented by Qual-Net simulator tools and by writing C++ programs, which are embedded with these tools.

8.1 Simulation environment

QualNet:- QualNet is a suite of tools to simulate enormous wired and wireless network systems. It supports to improve the performance of the network by creating an illusion of real-world scenario. It supports parallel computing and can run on multi-core systems, cluster systems, or multiprocessor systems, which make it capable to model huge networks with high fidelity. QualNet is compatible with various platforms, like Mac OS, Window XP, and Linux operating systems and supports 32- and 64-bit platforms. It provides the facility of graphical user interface and command line interface [38].

8.2 Network simulator parameters

The wormhole "all pass" model in QualNet network simulator (NS) was used to launch wormhole attack. During the simulation, many parameters were kept constant to observe better accuracy of results.

8.3 Simulation results

The simulation was carried out for duration 1000 s and repeated for 100 times for each parametric value. The results of the simulation are stored in the text file, and figures were plotted through MS Excel 2016. From the statistics file, we have computed average value Packet Delivery Ratio (PDR) and throughput for different scenarios. The scenario is verified within the same simulation environment by changing simulation parameters in Table 6 only.

8.3.1 Simulation-scenario 1

In the performance evaluation of the DSR, both DSR and TTWD under the wormhole attack were compared in terms of PDR and throughput by changing node mobility. The wormhole attack results in a steady reduction in PDR value as compared with the DSR. The TTWD model improved results as shown in Fig. 5.

Fig. 5 depicts maximum PDR growth (11%) at mobility of 35 m/s, while at the node mobility of 25 m/s, the PDR declined to approximately 6% only.

Table 6 Parameters changed during simulation.

S. no	Node density	Node mobility	Tunnel length	Malicious nodes
Scenario 1	50	5, 15, 25, 35, 45, 55, 65	2–8	2
Scenario 2	10, 20, 30, 40, 50	0–10	2–8	2
Scenario 3	50	0–10	2, 3, 4, 5, 6, 7, 8	2
Scenario 4	50	0–10	0–8	2

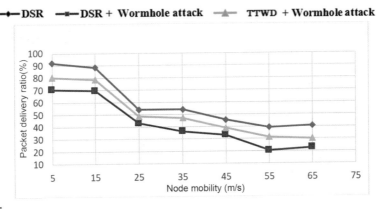

FIG. 5

Packet deliver ratio (PDR) verses node mobility.

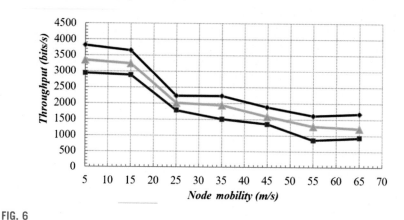

FIG. 6

Throughput versus node mobility.

The same pattern was observed for throughput when TTWD is applied. As shown in Fig. 6, TTWD showed that maximum growth in throughput is at speed of 35 m/s. It is approximately 20%, while minimum growth was at speed of 25 m/s, which is only 10%.

On increasing the speed of the nodes, the topology changes very rapidly. Due to this, building new routes become difficult and also the frequency of route breakage increased. Therefore the PDR and throughput decreases with increase in the speed of nodes. The PDR and throughput both initially fell rapidly. But with the further increment in speed, route breakage was going to be saturated. Thus both metric values drop comparatively low. It is observed that the TTWD technique performed better in the same simulation environment. It showed significant growth in PDR and throughput value. On average, in the presence of wormhole attack, TTWD achieved increment of 9% in PDR and 15% in throughput as compared with DSR.

8.3.2 Simulation-scenario 2

Under wormhole attack with different node density, the PDR, value, and throughput decreased as expected theoretically by 29% and 32% (on average), respectively. When TTWD model applied in presence of wormhole attack, then it achieved its best performance at node density 40 where TTWD model increased the PDR value by 16% as compared with DSR under wormhole attack as shown in the Fig. 7. At node density 10, TTWD showed its poorest performance where it increased PDR value only by 8% approximately.

Fig. 8 showed TTWD performance in term of throughput. The similar pattern matched with PDR as it showed best and worst performance on node density 40 and 10, respectively.

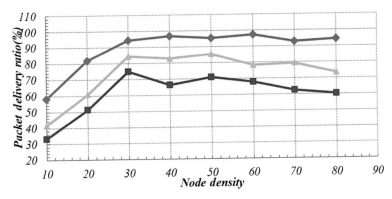

FIG. 7

Packet deliver ratio versus node density.

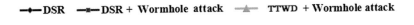

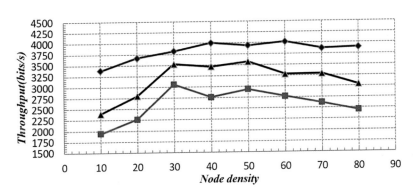

FIG. 8

Throughput verses node density.

The maximum growth in throughput is 17%, while the minimum is 13%. On average, TTWD achieved 13% increment in PDR, while 14% in throughput as compared with DSR under wormhole. Due to low node's density, that is, 10, there was connectivity problem because nodes that were spread in large area. Therefore nodes were not within the transmission range of each other. As the node density rose up, then it leads to better connectivity between the nodes. Therefore the PDR and throughput increased.

If the node density is further increased, connectivity between nodes improved, but more route breakage occurs due to large number of collisions. It also makes it difficult to establish new route due to collision. Thus the PDR and throughput showed a little up and down after node density 40.

8.3.3 Simulation-scenario 3

Two malicious nodes were chosen to launch the wormhole attack with different tunnel length. The TTWD model was applied to measure its detection rate. With the raise in tunnel length up to six hops, detection rate also increased exponentially. As depicted in Fig. 9, TTWD showed 100% detection on the tunnel length of six hops and more.

In the scenario without wormhole attack, the average RTT value between neighbors was 14 ms under TTWD model. The RTT between fake neighbors increased with increment in the hop length between them. Thus RRT value between fake neighbors exceeded the threshold value. The more the tunnel length, the more will be RTT value. Therefore wormhole attack detection rate improved with increment in the tunnel length.

8.3.4 Simulation-scenario 4

A detection of wormhole depends upon the comparison of RTT value with a threshold value. Therefore, to calculate the threshold limit, the value was chosen from RTT between actual neighbors and incremented further.

The higher threshold value with small tunnel length results large false negative, while low threshold value increased false positive. Fig. 10 displayed a high detection of wormhole after 25 s and achieved the highest accuracy at 35 s. At mobile of 35 milliseconds, both false negative and false positive were very low; therefore it was chosen as threshold limit value (Table 7).

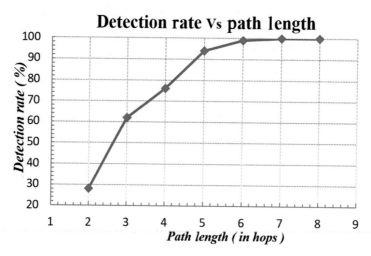

FIG. 9

Detection rate in TTWD.

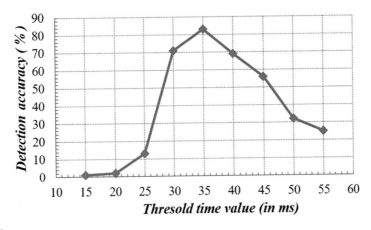

FIG. 10

Detection accuracy of TTWD.

Table 7 Comparison with other protocols.

Name	Based on	Extra hardware	Clock synchronization	Detection of wormhole node	Monitoring by neighbor
Packet Leashes	AODV	No	Yes	No	No
EDWA	AODV	Yes	Yes	Yes	No
WARP	AODV	No	No	Yes	Yes
WHOP	AODV	No	No	Yes	No
LITEWORP	DSR	Yes	Yes	Yes	Yes
DelPHI	AODV	No	No	No	No
TTWD	DSR	No	No	Yes	Yes

9 Conclusion and future directions

This chapter comprises abstract, introduction and motivation, the survey of security issues, threats and defense mechanisms, discussion over the efficiency of various counter measures, and the analysis of results obtained. In the future the proposed model can be enhanced by applying several machine learning algorithms for intrusion and malware detection in various communication networks. Further the proposed model can be implemented in real-world communication networks. Machine learning (ML) and IoT-based applications have taken the consideration of academicians, researchers, and industrialists in the last decade due to the constructive impact of these on human lives. IoT devices along with ML provide a way where small devices create a pool of relevant information, which is used by ML applications. Furthermore, ML applications produce complex visualizations, which are provided to IoT-based systems to facilitate the improvement of services and facilities. In the future, a supervised machine learning-based wormhole detection system (SMLDS) will be implemented so that system may detect a various types of attacks.

References

[1] K. Zhang, J. Ni, K. Yang, X. Liang, J. Ren, X.S. Shen, Security and privacy in smart city applications: challenges and solutions, IEEE Commun. Mag. 55 (1) (2017) 122–129.

[2] S. Ali, A. Ahmed, M. Raza, Towards better routing protocols for IoT, in: 2019 2nd International Conference on Computing, Mathematics and Engineering Technologies (iCoMET), IEEE, 2019, pp. 1–5.

[3] H. Bakht, Applications of Mobile Ad-Hoc Networks, Createspace Independent Publishing Platform, 2018.

[4] S.P. Khan, M.A. Rizvi, A.U. Mangal, Wireless ad hoc networks with 5G technology, in: Advanced Wireless Sensing Techniques for 5G Networks, Chapman and Hall/CRC, 2018, pp. 195–211.

[5] P. Kumar, N. Chauhan, N. Chand, Security framework for opportunistic networks, in: Progress in Intelligent Computing Techniques: Theory, Practice, and Applications, Springer, Singapore, 2018, pp. 465–471.

[6] M.M. Hossain, M. Fotouhi, R. Hasan, Towards an analysis of security issues, challenges, and open problems in the internet of things, in: 2015 IEEE World Congress on Services, IEEE, 2015, pp. 21–28.

[7] A. Singh, K.P. Kalita, S.P. Medhi, Blackhole attack on MANET and its effects, in: Proceedings of the 5th International Conference on Computing for Sustainable Global Development, New Delhi, India, 2018, pp. 14–16.

[8] V.S. Pooja, T. Rohit, N.M. Reddy, S. Sudeshna, Mobile ad-hoc networks security aspects in black hole attack, in: 2018 Second International Conference on Electronics, Communication and Aerospace Technology (ICECA), IEEE, 2018, pp. 26–30.

[9] A.K. Jain, V. Tokekar, Mitigating the effects of black hole attacks on AODV routing protocol in mobile ad hoc networks, in: 2015 International Conference on Pervasive Computing (ICPC), IEEE, 2015, pp. 1–6.

[10] A. Gaurav, A.K. Singh, Light weight approach for secure backbone construction for MANETs, J. King Saud Univ. Comput. Inf. Sci. (2018).

[11] M. Nawir, A. Amir, N. Yaakob, O.B. Lynn, Internet of things (IoT): taxonomy of security attacks, in: 2016 3rd International Conference on Electronic Design (ICED), IEEE, 2016, pp. 321–326.

[12] J. Swain, B.K. Pattanayak, B. Pati, Study and analysis of routing issues in MANET, in: 2017 International Conference on Inventive Communication and Computational Technologies (ICICCT), IEEE, 2017, pp. 506–509.

[13] R. Das, S. Bal, S. Das, M.K. Sarkar, D. Majumder, A. Chakraborty, K. Majumder, Performance analysis of various attacks under AODV in WSN & MANET using OPNET 14.5, in: 2016 IEEE 7th Annual Ubiquitous Computing, Electronics & Mobile Communication Conference (UEMCON), IEEE, 2016, pp. 1–9.

[14] S. Aluvala, K.R. Sekhar, D. Vodnala, An empirical study of routing attacks in mobile ad-hoc networks, Procedia Comput. Sci. 92 (2016) 554–561.

[15] A.S.K. Pathan (Ed.), Security of Self-Organizing Networks: MANET, WSN, WMN, VANET, CRC Press, 2016.

[16] S. Jamali, R. Fotohi, DAWA: defending against wormhole attack in MANETs by using fuzzy logic and artificial immune system, J. Supercomput. 73 (12) (2017) 5173–5196.

[17] A.T. Kolade, M.F. Zuhairi, E. Yafi, C.L. Zheng, Performance analysis of black hole attack in MANET, in: Proceedings of the 11th International Conference on Ubiquitous Information Management and Communication, ACM, 2017, p. 1.

[18] N. Yadav, V. Parashar, Trust or reputation base encryption decryption technique for preventing network from DOS attack in MANET, in: 2016 International Conference on Inventive Computation Technologies (ICICT), vol. 1, IEEE, 2016, pp. 1–6.

[19] G. Cerullo, G. Mazzeo, G. Papale, B. Ragucci, L. Sgaglione, IoT and sensor networks security, in: Security and Resilience in Intelligent Data-Centric Systems and Communication Networks, Academic Press, 2018, pp. 77–101.

[20] S. Deshmukh-Bhosale, S.S. Sonavane, A real-time intrusion detection system for wormhole attack in the RPL based internet of things, Procedia Manuf. 32 (2019) 840–847.

[21] M.A. Khan, K. Salah, IoT security: review, blockchain solutions, and open challenges, Futur. Gener. Comput. Syst. 82 (2018) 395–411.

[22] D. Airehrour, J.A. Gutierrez, S.K. Ray, SecTrust-RPL: a secure trust-aware RPL routing protocol for internet of things, Futur. Gener. Comput. Syst. 93 (2019) 860–876.

[23] A. Machanavajjhala, X. He, M. Hay, Differential privacy in the wild: A tutorial on current practices & open challenges, in: Proceedings of the 2017 ACM International Conference on Management of Data, ACM, 2017, pp. 1727–1730.

[24] D.S.K. Tiruvakadu, V. Pallapa, Confirmation of wormhole attack in MANETs using honeypot, Comput. Secur. 76 (2018) 32–49.

[25] A.K. Das, S. Zeadally, D. He, Taxonomy and analysis of security protocols for internet of things, Futur. Gener. Comput. Syst. 89 (2018) 110–125.

[26] P.I.R. Grammatikis, P.G. Sarigiannidis, I.D. Moscholios, Securing the internet of things: challenges, threats and solutions, Internet Things 5 (2019) 41–70.

[27] D. Giri, S. Borah, R. Pradhan, Approaches and measures to detect wormhole attack in wireless sensor networks: a survey, in: Advances in Communication, Devices and Networking, Springer, Singapore, 2018, pp. 855–864.

[28] M.S. Ahsan, M.N.M. Bhutta, M. Maqsood, Wormhole attack detection in routing protocol for low power lossy networks, in: 2017 International Conference on Information and Communication Technologies (ICICT), IEEE, 2017, pp. 58–67.

[29] I. Butun, P. Österberg, H. Song, Security of the Internet of Things: Vulnerabilities, Attacks and Countermeasures, IEEE Communications Surveys & Tutorials, 2019.

[30] S. Palacharla, M. Chandan, K. GnanaSuryaTeja, G. Varshitha, Wormhole attack: a major security concern in internet of things (Iot), Int. J. Eng. Technol. 7 (3.27) (2018) 147–150.

[31] S. Qazi, R. Raad, Y. Mu, W. Susilo, Multirate DelPHI to secure multirate ad hoc networks against wormhole attacks, J. Inf. Secur. Appl. 39 (2018) 31–40.

[32] F.A. Khan, M. Imran, H. Abbas, M.H. Durad, A detection and prevention system against collaborative attacks in mobile ad hoc networks, Futur. Gener. Comput. Syst. 68 (2017) 416–427.

[33] X. Luo, Y. Chen, M. Li, Q. Luo, K. Xue, S. Liu, L. Chen, CREDND: a novel secure neighbor discovery algorithm for wormhole attack, IEEE Access 7 (2019) 18194–18205.

[34] S. Gupta, S. Kar, S. Dharmaraja, WHOP: Wormhole attack detection protocol using hound packet, in: Proc. IEEE on Innovations in Information Technology, 2011, pp. 226–231.

[35] B. Bhushan, G. Sahoo, Recent advances in attacks, technical challenges, vulnerabilities and their countermeasures in wireless sensor networks, Wirel. Pers. Commun. 98 (2) (2018) 2037–2077.

[36] M. Rmayti, Y. Begriche, R. Khatoun, L. Khoukhi, A. Mammeri, Graph-based wormhole attack detection in mobile ad hoc networks (manets), in: 2018 Fourth International Conference on Mobile and Secure Services (MobiSecServ), IEEE, 2018, pp. 1–6.

[37] A.K. Yadav, A. Kush, TCP-and UDP-based performance evaluation of AODV and DSR routing protocol on varying speed and pause time in mobile ad hoc networks, in: Next-Generation Networks, Springer, Singapore, 2018, pp. 323–332.

[38] M. Dhingra, S.C. Jain, R.S. Jadon, Performance comparison of LANMAR and AODV in heterogenous wireless ad-hoc network, in: Emerging Trends in Expert Applications and Security, Springer, Singapore, 2019, pp. 125–132.

Fault tolerance of cluster-based nodes in IoT sensor networks with periodic mode of operation

7

Igor Kabashkin

Transport and Telecommunication Institute, Riga, Latvia

Chapter outline

1 Introduction

Wireless sensor networks (WSNs) are being actively developed at present. They are one of the basic technologies using various IoT applications especially in cyber-physical systems (CFS). Such applications integrate different technologies and include different network capabilities [1].

In WSN, many interconnected sensor (S) nodes with wireless channels form a spatially distributed system. In large networks, sensors have restrictions on the speed and amount of information processed. To reduce these restrictions on large spatial areas, sensors in networks can be aggregated in spatially distributed groups called clusters. The creation of these clusters can significantly improve the efficiency of both sensors in groups and the entire network as a whole. The data collected by the sensors of each node are transmitted to the central element of the cluster, which acts as the cluster head element (CH). The base station (BS) collects information

133

Security and privacy issues in IoT devices and sensor networks. https://doi.org/10.1016/B978-0-12-821255-4.00007-9

from all clusters through their head elements. A cluster-oriented structure of WSN is shown at the Fig. 1.

The cyber-physical system is usually designed for autonomous functioning without direct participation and control by humans. CFS is often designed to collect information and monitor the status of highly responsible and mission critical systems.

Dependability and fault tolerance are important attributes of such WSNs. In these networks, requirements for the reliability of the network as a whole and the availability of individual sensor nodes are dominant. In a number of works, the issues of the influence on the reliability of networks of their topology, protocols, and application-level error correction are investigated [2, 3].

Current studies of the reliability of the sensor node mainly investigate the effect on the network efficiency of problems related to equipment reliability and communication problems between network nodes [4]. Sensors usually have autonomous power supply from accumulators or batteries, which is one of the critical factors in the life cycle of a network and requires additional attention from the point of view of its fault tolerance. For sensors that work in wireless networks, the power supply is one of the key elements. Since the sensors are distributed, a typical energy supply structure for them is an individual energy supply using a personal battery (Fig. 2).

In real cyber-physical systems, sensors usually work periodically, fixing discrete changes in the monitoring objects. The periods of sensor activity t_{ms} during operation of sensors functioning in the network alternate with sleep mode T_{ms} (Fig. 3). For systems with a similar periodic nature of functioning, there is a probability of a system failure in the period between active works, which will not be fixed. In this case the

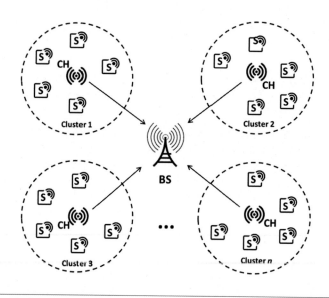

FIG. 1

Architecture of cluster-oriented WSN.

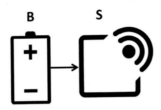

FIG. 2

Traditional sensor with supported battery.

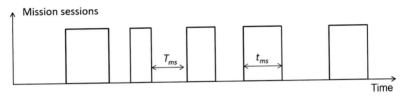

FIG. 3

Operation of sensor during life cycle.

next period of active work will be in a failed state, which will lead to the inability of the sensor to perform its function.

A similar situation can arise in wireless networks of the pulsed nature of the sensors work, in which, with a relatively low average battery power consumption, pulse surges during periods of information transfer in communication channels can be quite high [5]. In this case time t_{ms} is active mode duration (in milliseconds range); time T_{ms} is sleep or pause mode duration (this time for various systems can range from seconds to hours). During periods of inactive operation of the sensors, there is no need to work on transmission in the communication channel. At this time the battery is not actively used and gradually restores the lost capacity, which was pulse, spent in active mode. In a situation where there is not enough time for relaxation, the limited capacity of the batteries can cause a sensor failure, even if its average energy consumption is not so large [5]. When the battery level drops below a critical level, the sensor cannot perform its functions. In this case the life cycle of the operation of cyber-physical systems can be extended by introducing the redundancy of the power supply of wireless network sensors due to battery backup [6]. Using this method, each of the m sensors works with a pair of batteries, one of which is in the main mode, and the second performs the backup function. The battery communication switch (BCS) performs the functions of monitoring the battery capacity and switching them to standby (Fig. 4).

The paper analyzes the dependability of the cluster-oriented nodes in IoT wireless networks with periodic active mode of sensors operation with built-in test diagnostics systems (DS). The sensors power sources have a limited operating time, which is

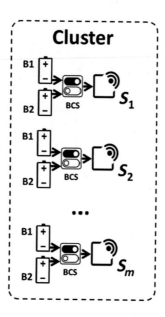

FIG. 4

Cluster of sensors with duplication set of batteries.

due to their physical nature of work and the intensity of information exchange in a particular wireless network. The resiliency of WSN in such systems may be signif-icantly increased by increasing the time of uninterrupted power supply of wireless sensor networks. To achieve this goal, this paper describes the method of redundancy of batteries for sensor nodes in cluster-based WSN. In contrast to the duplication of the battery of each sensor in a cluster-oriented network, it becomes possible to use a common set of backup batteries as an additional method of fault tolerance. These excess batteries can be used to maintain the availability of any of the sensors in the event of a failure of its individual batteries.

The chapter has the following structure. Review of references in the field of dependability of sensor nodes in WSN is carried out in Section 2. The basic defini-tions and notation used in the chapter are given in Section 3. Section 4 describes the proposed approaches to improve fault tolerance of WSN. In the same section, models for the reliability of sensors in cluster-oriented WSN CPS are developed, and their comparative analysis is performed. Conclusions are given in the Section 5.

2 Related works

In the paper [7] an overview of existing protocols for the reliability of data transmis-sion in WSN is presented. The paper considers several reliability models. They are focused on the use of various mechanisms for recovering lost data with insufficient reliability of packet transmission, in particular, through the use of retransmission methods and coding redundancy.

Survey of different dependability methods in WSN and specific sensors is presented in paper [8].

Main of the references on the dependability analysis of WSN uses graph theory and probability theory. Based on the application of these theories in publications [9–12], the network fault tolerance factor is determined as the probability of preserving a certain functionality with partial failures of some channels and nodes. The papers [13, 14] study delay effects, speed in the network, and other network parameters with different node reliability. A classification of reliability and security factors was defined in [15] for selection of the cluster head and description of clustering in WSN.

Simulation is often used for analyzing the dependability of networks. This is determined by the complexity of real influence of various factors on network reliability (hardware and software failures, environmental influences, and others) and by difficulty to describe it mathematically. These factors and network behavior from dependability point of view are described by Bayesian networks [16], Petri nets [17], and Monte Carlo modeling [18].

Reliability of wireless networks is largely determined by the reliability of sensors. In [19] the sensor was considered at the microlevel; its components providing communication are analyzed for influence on the reliability of WSN as a whole. Reliability of network sensors and methods for increasing them are carried out using probabilistic analysis based on Markov models.

In the real world of using wireless sensor networks, various maintenance methods are used to ensure their reliable operation. The influence of the frequency of maintenance on the availability of network nodes, as well as the optimal conditions for its implementation, is studied in [20].

In article [21] a malfunction or a compromised node is studied. Under these conditions the head node may receive false information due to data corruption. Data aggregation in this case becomes problematic, as it becomes a factor of influence on network reliability.

The impact of power outages is especially sensitive in systems critical to safety. New generations of autonomous objects with built-in redundancy of individual subsystems and components, including energy supply sources, are considered in [22].

Electric cars and their functioning are a new area that is intensifying research on the reliability of energy supply in network structures. The approach to ensuring the fault tolerance of power supply of electric vehicles based on dynamic switching of the battery network and their optimal balancing is considered in [23].

Troubleshooting of wireless sensor networks is a separate area of research. A review of studies in this direction is contained in [24]. The study provides a classification of diagnostic methods by type of test, the nature of the test signals, and the test methodology.

Duplication of power supply sources of individual nodes of wireless sensor networks as a method of increasing reliability is considered in [6]. For the proposed method a model of the dependability of the sensor node is proposed, and investigation of its availability in comparison with the traditional method of using batteries is made.

The use of duplication of power sources of individual sensor nodes increases their reliability but at the same time increases the resources to ensure the reliability of their functioning. In [25] an approach was proposed that allows one to find a compromise between these factors based on the use for cluster sensors of a one mutual set of backup energy suppliers. A network reliability model is obtained to determine the availability of the sensor node using the proposed approach. Such method is known as *k-out-of-n* system structure. It is used as method of increase resilience in the systems with uniform elements. The method *k-out-of-n:G* [26–28] describes system with *n* identical elements. The system will fail if more than *n-k* of the elements in the system will fail. The similar model of *k-out-of-n:F* redundancy [29] is a system with *n* identical elements, but the system will fail if more than *k* of the elements in the system will fail. Applications of *k-out-of-n* structure are very popular for design of fault tolerant telecommunication systems [30].

In real systems, which are critical for security, not only the reliability of the network as a whole but also the availability of individual channels for receiving information is important from the point of view of fault tolerance. The influence of the used backup methods in such systems was studied in [31]. The reliability of switching functions in a redundant switching device is of particular importance in such redundant systems. The architecture of such automatic switching devices allows the possibility of switching failures of two types—"false switching" and "without switching." In [25], models of system reliability for both types of switch failures are considered, and requirements for the probability of each type of failure for highly reliable systems are determined.

One of the criteria of WSN reliability is resilience for network topology changing and congestion issues. In the [32], clustering and routing protocol to alleviate the congestion issue over the network are proposed. Some theoretical and practical results to decrease end-to-end delay time and prolong the network lifetime through choosing the suitable primary cluster head and the secondary cluster head are discussed in the paper.

Additional factor of reliability is protection information in sensor network against various external attacks that may make the WSN systems vulnerable and unstable. The paper [33] studies the reliability of WNS from the security point of view imposed by the open nature of networks. The paper provides a detailed survey of data aggregation and the energy-efficient routing protocols for WSN and define few unsolved technical challenges and the future scope for WSN security.

Also, one aspect of dependability of WSN connected with drawback of the clustering approach is the imposed energy overhead caused by the global clustering operations in every round of the global round-based policy. To mitigate this problem the paper [34] proposes a hierarchical clustering-task scheduling policy, which triggers node-driven clustering as opposed to global round-based policy time-driven clustering.

Sensor nodes in WSN located in sinks vicinity deplete their batteries quickly because of concentrated data traffic near the sink, leaving the data reporting wrecked and disrupted. One of the solutions of this problem may be mobile sinks that are introduced and provide uniform energy consumption and load-balanced data delivery through the sensor network [35]. An energy-efficient distributed mobile sink routing protocol has been proposed in [36] that considers rechargeable sensors to be deployed in the sensing region and employs maximum capacity path, a dynamic load balanced routing scheme for load balancing and prolonging the networks lifetime. Providing the innovative solutions to reduce the congestion issue over the network has become necessary not only to decrease network bandwidth and power usage but also to prolong the network lifetime as much as possible [37].

The survey in the paper [38] classifies fault diagnosis methods in recent 5 years into three categories based on decision centers and key attributes of employed algorithms: centralized approaches, distributed approaches, and hybrid approaches.

In this chapter the reliability of a single sensor (Fig. 1) with periodic active modes (Fig. 2) and various battery testing strategies is considered. As an additional tool for improving fault tolerance in WSN cluster nodes, there is a single set of backup batteries, each of which can be used to provide power to any of the sensors in the cluster (Fig. 5). The system redundancy architecture provides for each sensor in the cluster a primary energy source (main battery [MB]) and a common set of backup energy sources (redundant batteries [RB]). The matrix communication switch (MCS) replaces any failed battery with a functioning backup battery.

We will build a model of sensor reliability in a cluster-oriented network for the described architecture of battery backup in real-world conditions with periodic operation and two possible strategies for the operation of DS.

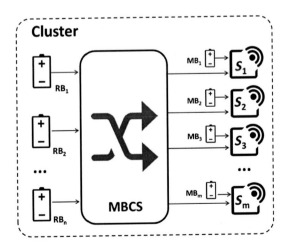

FIG. 5

Sensor cluster with common pool of standby batteries.

3 Mathematical background and main symbols and definitions

In this chapter the fault tolerance of cluster-based nodes in IoT sensor networks with periodic mode of operation is investigated. The analysis is performed on the basis of modeling these systems using Markov models of reliability based on the classical mathematical models with construction of Kolmogorov-Chapman differential equations and determining the stationary values of the probabilities of finding the system under study in a stationary state [39].

Symbols and notation used in the study for the development of reliability models are shown in Table 1.

The reliability of all switching elements in the system is considered ideal; all time intervals have exponential distribution.

Table 1 The symbols used for the design and analysis of mathematical models.

Symbols	Definitions
λ	Failure rate of all batteries
μ	Repair rate of all batteries
A_0	Availability of the ideal sensor without test operations
A_1	Sensor availability with first test strategy
A_2	Sensor availability with second test strategy
A_{ss}	Availability of sensor in cluster-based nodes of WSN
V	Availability degradation factor
T_0	Mean time between failures, $T_0 = 1/\lambda$
m	Number of sensors in cluster
n	Number of mutual set of redundant batteries in cluster
p_i	Probability for the state H_i
T_m	Periodicity of test operations with parameter $\omega = 1/T_m$
t_m	Duration of test operations
τ	Parameter of exponential distribution of t_m
t_s	Mean time of backup battery switching on a reserve battery
$\upsilon = 1/t_s$	Parameter of t_s distribution
t_f	Mean time of failure detection by diagnostics
$\nu_1 = 1/t_f$	Parameter of t_f distribution
T_c	Mean time between active modes, $\varphi = 1/T_c$
t_c	Mean time of active mode for sensor
$\psi = 1/t_c$	Parameter of t_c distribution
t_{ti}	Mean time of diagnostics interruption if sensor start operation in active mode
$\nu_2 = 1/t_{ti}$	Parameter of t_{ti} distribution

4 Models formulation and solution

Embedded test equipment is a typical element of the architecture of modern systems [24]. In systems with a discrete pulse mode of operation of sensors (Fig. 2), two strategies for monitoring its operability are possible: testing in the active mode of operation (strategy 1) and testing in the active mode of operation with additional periodic diagnostics in the mode of operation pause (strategy 2). Sensor reliability models for each of their diagnostic strategies are given in the succeeding text.

4.1 Model 1. Sensor with diagnostics in active mode

The Markov model can be used to determine the reliability of the system under study. The transition diagram for states of the studied system is shown in the Fig. 6, where H_1, completely good state of sensor; H_2, active mode for sensor without battery failure; H_3, failure of the sensor's battery; H_4, active mode for sensor with battery failure; and H_5, detection of the failure state.

We can write the system of Chapman-Kolmogorov's equations of Markov model for the state transition diagram shown in Fig. 6:

$$P_1'(t) = \psi P_2(t) - (\varphi + \lambda)P_1(t) + \mu P_5(t)$$

$$P_2'(t) - \varphi P_1(t) - (\psi + \lambda)P_2(t)$$

$$P_3'(t) = \lambda P_1(t) - \varphi P_3(t)$$

$$P_4'(t) = \lambda P_2(t) + \varphi P_3(t) - v_1 P_4(t)$$

$$P_5'(t) = v_1 P_4(t) - \mu P_5(t)$$

In the aforementioned equations, $P_i(t)$ is the probability of state $H_i(t), i = \overline{1,5}$.

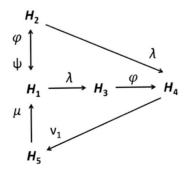

FIG. 6

State transition diagram for model 1.

The normalization condition is

$$\sum_{i=1}^{5} P_i(t) = 1$$

The availability of the system can be defined as $A_1 = P_1 + P_2$ (P_i, stationary probability of the state $H_i(t), i = \overline{1,n}$) of the sensor with the first model of diagnostics that may be defined from aforementioned system of Chapman-Kolmogorov's equations:

$$A_1 = (1 + a_1)/(1 + a_1 + a_2), \tag{1}$$

$$a_1 = \lambda \left[\gamma(1+\beta) + \frac{1}{\varphi} \right], a_2 = \frac{\varphi}{\lambda + \psi}, \gamma = \frac{1}{\mu} + \frac{1}{v_1}$$

Availability A_1 of sensor with the first diagnostics model can be compared with the availability A_0 of the ideal sensor dependability [39]. Comparison shows that $A_1 = A_0$ with decreasing of the mean time between active modes of sensors T_c and with the increasing the parameter of distribution for time of failure detection by diagnostics $v_1 \to \infty$.

4.2 Model 2. Sensor with diagnostics in active mode and periodical diagnostics in the sleep mode

The reliability of this system can be described by Markov model with transition diagram of state, shown at the Fig. 7, where H_1, good condition of sensor, absence of the active mode, and diagnostics operations; H_2, test in the sensor with inactive mode of operation; H_3, active mode of operation; H_4, demand on active mode operation in the test period; H_5, failure of battery in the sleep mode; H_6, test of sensor with battery failure; H_7, demand on active mode in the sensor with battery failure; and H_8, fixation of failure by built-in test system.

The Chapman-Kolmogorov's equations for aforementioned state transition diagram (Fig. 7) can be written as next system of equations:

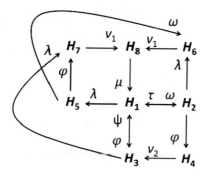

FIG. 7

State transition diagram for model 2.

$$P'_1(t) = \tau P_2(t) - (\varphi + \lambda + \omega)P_1(t) + \psi P_3(t) + \mu P_8(t)$$

$$P'_2(t) = \omega P_1(t) - (\varphi + \tau + \lambda)P_2(t)$$

$$P'_3(t) = \varphi P_1(t) - (\lambda + \psi)P_2(t) + \upsilon_2 P_4(t)$$

$$P'_4(t) = \varphi P_2(t) - \upsilon_2 P_4(t)$$

$$P'_5(t) = \lambda P_1(t) - (\omega + \varphi)P_5(t)$$

$$P'_6(t) = \lambda P_2(t) + \omega P_5(t) - \upsilon_1 P_6(t)$$

$$P'_7(t) = \lambda P_3(t) + \varphi P_5(t) - \upsilon_1 P_7(t)$$

$$P'_8(t) = \upsilon_1 P_6(t) + \upsilon_1 P_7(t) - \mu P_8(t)$$

For aforementioned system of equations, the normalizing condition is

$$\sum_{i=1}^{8} P_i(t) = 1$$

The dependability of the sensor with the second model of diagnostics can be described by availability $A_2 = P_1 + P_2 + P_3$ that can be determined in a standard way by solving of the indicated Chapman-Kolmogorov's equations:

$$A_2 = (1 + a_2)/(1 + a_1 + a_2) \tag{2}$$

$$a_1 = \omega/(\lambda + \varphi + \tau)[\lambda(1/\mu + 1/\upsilon_1) + \varphi/\upsilon_2 + \lambda\varphi(1/\mu + 1/\upsilon_1)/(\lambda + \psi)]$$
$$+ \lambda/(\omega + \varphi)[1 + \omega(1/\mu + 1/\upsilon_1) + \varphi(1/\mu + 1/\upsilon_1)] + \varphi(1 + \lambda/\upsilon_1)/(\lambda + \psi),$$

$$a_2 = (\lambda + \psi)^{-1}[\omega(\lambda + \varphi + \psi)/(\lambda + \varphi + \tau) + \varphi].$$

The analysis of the function $A_2(\omega)$ shows that it has extremes in ω_{opt} point.

The optimal periodicity of diagnostics $T_{m\ opt} = 1/\omega_{opt}$ can be find by solution of the equation $dA_2/d\omega = 0$. The mathematical expression of the optimal periodicity of diagnostics $T_{m\ opt}$ has rather complicated expression in general case. In practice the batteries of sensors have high dependability ($\lambda \ll \mu$) with the quick time of failure detection by diagnostics ($\mu \ll \upsilon_1$, $\mu \ll \upsilon_2$). In this case the mathematical expression for optimal periodicity of diagnostics can be defined by the next formula:

$$T_{m\ opt}^{-1} = \upsilon_2 \frac{\left[\lambda + \sqrt{\left(\lambda - \frac{\varphi^2}{\upsilon_2}\right)^2 + \frac{\lambda\varphi\tau}{\upsilon_2}}\right]}{\varphi} - \varphi$$

4.2.1 Numerical example

The dependability of sensor nodes in WSN can be evaluated by the availability of sensor in the node with periodical active modes of operation. The moment of the battery failure could be determined by the periodical self-diagnostics in the sleep mode of sensor.

Let us compare the reliability of real sensor with the ideal one by the availability degradation factor:

$$V = \frac{1-A_i}{1-A_0}, i = 1,2$$

At Fig. 8 the degradation factor as function of periodicity of sensor active modes T_c is shown for the first model of diagnostics with mean time between failure of batteries 3000 h:

$$V = \frac{1-A_1}{1-A_0}$$

The availability A_1 is calculated by Eq. (1); expression for the availability A_0 of the ideal elements was determined in [39].

At Fig. 9 the degradation factor is shown as function of periodicity of diagnostics T_m for various mean time between failure T_0:

$$V = \frac{1-A_2}{1-A_0}$$

The availability A_2 is calculated by Eq. (2); expression for the availability A_0 of the ideal elements was determined in [39].

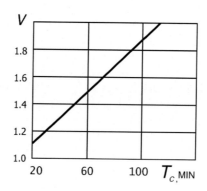

FIG. 8

The reliability degradation factor of sensor for model 1.

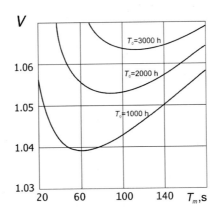

FIG. 9

The reliability degradation factor of the sensor with model 2.

4.3 Model 3. Sensors with mix architecture of backup batteries

Duplication of batteries for sensors (Fig. 4) is one of the methods to increase the fault tolerance of sensors in wireless networks. With this backup architecture, each network sensor has a duplicate set of batteries that are switched by a matrix battery communication switch (MBCS).

To increase the fault tolerance of network nodes, we use a backup architecture in which we additionally apply a common backup set of batteries for energy supply reservation of all sensors in cluster (Fig. 10).

We investigate a WSN with $N = m + n$ batteries in cluster, m of which are main batteries and n are standby batteries used as a common redundant set of elements (Fig. 10). All reconfigurations of redundant architecture are performed by MBCS. Dependability of the MBCS in comparison with the dependability of the batteries will be considered ideal. The availability A_{ss} of any selected sensor in WSN cluster should be no worse when using the first and second diagnostic models

$$A_i \geq A_{ss}, i = 1, 2 \tag{3}$$

Markov model of reliability for studied system with $1 \leq l \leq n$ repair bodies and active standby mode of redundant batteries can be described by the state transition diagram shown at Fig. 10, which has the next states: H_i, cluster has i batteries with failure, and selected sensor has a working battery; H_{il}, cluster has $i + l$ batteries with failure, and selected sensor has battery with failure; H_{Ti}, cluster has i batteries with failure, $i - 1$ battery are in recovery mode, failure of ith battery is not detected, sensor has a demand on active mode, and selected sensor has a working battery; $H_{\pi i}$, cluster has i batteries with failure, $i - 1$ batteries are in recovery mode, failed battery is under replacement by MBCS, and selected sensor has a working battery; H_{ITi}, i failed batteries, selected sensor has battery with failure that is not detected, sensor has a demand on active mode, and the $i - 1$ failed batteries are in recovery mode; and $H_{I\pi i}$,

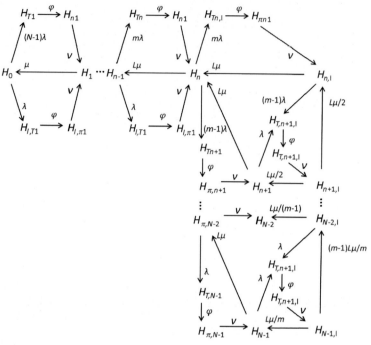

FIG. 10

State transition diagram for model 3.

i failed batteries, selected sensor has battery with failure, failed battery is under replacement by MBCS, and $i-1$ batteries are in recovery mode.

The Markov model of reliability with the state transition diagram (Fig. 10) can be described by the system of Chapman-Kolmogorov's equations written in accordance with the general rules [39].

The expression for the availability of the sensor in the cluster of the wireless sensor network can be determined in a standard way:

$$A_{ss} = 1 - \sum_{\forall i,j} P_{ijl} = \frac{a_1 + a_2}{a_1 + a_2 + a_3},$$

$$a_1 = \sum_{i=0}^{l} \binom{i}{N} \gamma^i \left[1 + \lambda \left(\frac{1}{\varphi} + \frac{1}{v}\right)(N-i-1)\right] + \frac{N!l^l}{l!} \sum_{i=l+1}^{n-1} \frac{\omega^i}{(N-i)!} \left[1 + \lambda \left(\frac{1}{\varphi} + \frac{1}{v}\right)(N-i-1)\right],$$

$$a_2 = \frac{N!l^l \omega^n}{ml!} \sum_{i=0}^{m-1} \frac{\omega^i}{(m-i-1)!} \left[1 + \lambda \left(\frac{1}{\varphi} + \frac{1}{v}\right)(m-i-1)\right],$$

$$a_3 = \frac{N! l^l \omega^n}{ml!} \sum_{i=0}^{m-1} \frac{(i+1)\omega^{i+1}}{(m-i-1)!} \left[1 + \lambda \left(\frac{1}{\varphi} + \frac{1}{\upsilon} \right) \left(m - i - 1 + \frac{1}{i+1} \right) \right]$$

$$+ \lambda \left(\frac{1}{\varphi} + \frac{1}{\upsilon} \right) \left[\sum_{i=0}^{l} \binom{i}{N} \gamma^i + \frac{N! l^l}{l!} \sum_{i=l+1}^{n-1} \frac{\omega^i}{(N-i)!} \right],$$

$$\gamma = \frac{\lambda}{\mu}, \omega = \frac{\gamma}{l}.$$

The node cluster of sensors in WSN for IoT usually is a system with high dependability. In this case inequality $N\omega \ll 1$ is correct, and expression for availability of selected sensor in the cluster A_{ss} can be written with enough accuracy for practical purposes as

$$A_{ss} = 1 - \lambda \left(\frac{1}{\varphi} + \frac{1}{\upsilon} \right) a, \tag{4}$$

$$a = \sum_{i=0}^{l} \binom{i}{N} \gamma^i + \frac{N! l^l}{l!} \sum_{i=l+1}^{n} \frac{\omega^i}{(N-i)!}.$$

Limitation on switch time $t_\varsigma = 1/\upsilon$ within the frame of standby commutation can be determined by substituting the Eq. (4) in the formula (3):

$$t_s \leq \frac{1 - A_0}{a\lambda} - \frac{1}{\varphi} \tag{5}$$

The analysis of reliability for the WSN with $n \leq l \leq m$ repair bodies is shown that if case inequality $N\omega \ll 1$ is correct, the limitation on switch time $t_s = 1/\upsilon$ during reservation process can be calculated by the same formula (5), where

$$a = \sum_{i=0}^{n} \binom{i}{N} \gamma^i$$

5 Results and discussions

In the previous section the reliability models of cluster-based nodes in IoT sensor networks with periodic mode of operation are developed. The analysis of their dependability is performed on the basis of modeling this system using Markov models of reliability based on the classical mathematical models with construction of Kolmogorov-Chapman differential equations and determining the stationary values of the probabilities of finding the system under study in a stationary state.

For a discussion of the impact of various factors on the reliability of the studied systems, we consider several numerical examples.

The reliability of sensor cluster can be analyzed with availability of selected sensor in cluster-oriented wireless sensor network in accordance with the mix architecture of backup batteries of model 3.

Evaluation of dependability in the studied sensor cluster with mix architecture of backup batteries (model 3) in comparison with the standard model of duplicate battery with two strategies of diagnostics operations (models 1 and 2) can be carried out with the availability degradation factor:

$$V = \frac{1 - A_d}{1 - A_{ss}}$$

The value of availability of selected sensor in the cluster A_{ss} can be determined by the Eq. (4); formula for the availability of the system with duplicate elements A_d can be borrowed from [39].

The degradation factor V at Fig. 11 is shown as function of number m sensors in cluster of wireless sensor network with different number n backup batteries in common standby pool for typical reliability parameters of each battery (mean time between failure of batteries is 3000 h; mean time between active modes is 30 min).

The results of the analysis of the graphs in Fig. 11 show that for WSN with cluster-based architecture and small size of cluster (number of sensors $m \leq 8$) with proposed architecture of batteries redundancy, the sensor dependability is not less than in the architecture with the duplication of RB and is achieved by using only two redundant batteries in common pool.

For the networks with $m \geq 9$ sensors in cluster, application with common set of standby batteries requires only three batteries in common pool for the same sensor reliability. In this case, for network with $m = 8$ sensor in node, the retrenchment will be six free batteries, and for $m = 21$ sensors, this effect will be 18 free batteries.

Fig. 12 shows the dependence of the maximum allowable switching time to reserve t_s on the average time T_{ms} of the sensor staying in the active mode. The dependency is built for WSN cluster with $n = 10$ sensors for the condition of equal reliability of the sensor in the cluster for both architectures ensuring the reliability of

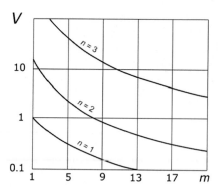

FIG. 11

The reliability degradation factor.

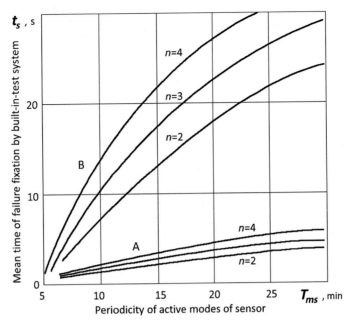

FIG. 12

Requirements for the time to switch batteries to standby (A, for first test strategy; B, for second test strategy).

energy supply: duplication of batteries (Fig. 4) and backup with a common set of batteries (Fig. 5). Curves A at the Fig. 12 show the indicated dependence for the first test strategy of sensors (model 1) and curves B—for the second test strategy of sensors (model 2). In reality, Fig. 12 represents lines of equal reliability, the points of which satisfy the condition $A_{ss} = A_d$ and correspond to the critical values t_s of the switching time to the reserve. In this case the environment lying below the curves corresponds to the case where $A_{ss} > A_d$, and the environment lying above the curves corresponds to the case $A_{ss} < A_d$.

An analysis of dependencies shown in Fig. 12 leads to the conclusion that when there are $n > 2$ backup batteries in a system with an additional common set of backup batteries, there is always such a switchover time to a reserve t_s at which the sensor reliability in backup architecture with a common set of batteries (Fig. 5) will always be higher than in architecture with duplication of batteries (Fig. 4).

6 Conclusions

Periodical operation of sensors for different type of WSNs is typical for cyber-physical system. The paper analyzes different methods for increasing fault tolerance and autonomy time of sensors with periodic mode of operation in cluster-based WSN with special accent on sensor's battery power efficiency.

The article shows that the reliability of information transfer by sensors in such systems is sensitive to the testing strategy. Among alternative strategies of the sensor health monitoring, test strategy with diagnosis during active mode of operation and additional periodical test in the sleep mode of sensor functionality is about twice more reliable than test strategy of sensors with health monitoring only during active mode of operation.

The availability of the sensors with health monitoring during active mode of operation has optimal periodicity of diagnostics operations. The optimal periodicity of diagnostics will increase than mean time between failures increase. The same situation will be in case than time of diagnostics, and time of its interruption is increased.

The optimal periodicity of diagnostics is not depending from the duration of the active mode of sensor operation. The expression for optimal periodicity of diagnostics is determined.

As an additional method of fault tolerance in cluster-based nodes of WSN with periodic operations, the possibility of using an additional set of batteries in a cluster is considered. These redundant batteries can be used to maintain the availability of any of the sensors in the case of failure of his individual batteries.

It is shown that when there are $n > 2$ backup batteries in a system with an additional common set of backup batteries, there is always such a switchover time t_s at which the sensor reliability in backup architecture with a common set of batteries will always be higher than in architecture with batteries duplication.

Markov models of reliability analyses of sensor cluster in wireless sensor network with proposed methods of fault tolerance are developed. Expressions for the availability of sensors in the described network and for the fault tolerance effect of network operation are obtained.

References

[1] G.J. Pottie, Wireless integrated network sensors (WINS): the web gets physical, in: Frontiers of Engineering: Reports on Leading-Edge Engineering from the 2001 NAE Symposium on Frontiers of Engineering, National Academies Press, 2002, p. 78.

[2] A. Ayadi, Energy-efficient and reliable transport protocols for wireless sensor networks: state-of-art, Wirel. Sens. Netw. 3 (3) (2011) 106–113, https://doi.org/10.4236/wsn.2011.33011.

[3] K. Sharma, R. Patel, H. Singh, A reliable and energy efficient transport protocol for wireless sensor networks, Int. J. Comput. Netw. Commun. 5 (2) (2010) 92–103.

[4] S.-J. Park, R. Sivakumar, I.F. Akyildiz, et al., GARUDA: achieving effective reliability for downstream communication in wireless sensor networks, IEEE Trans. Mob. Comput. 7 (2) (2008) 214–230, https://doi.org/10.1109/TMC.2007.70707.

[5] S. Farahani, Battery life analysis, in: ZigBee Wireless Networks and Transceivers, 2008, pp. 207–224.

[6] S. Mahajan, P. Dhiman, Clustering in wireless sensor networks: a review, Int. J. Adv. Res. Comput. Sci. 7 (3) (2016) 198–201.

[7] M. Mahmood, W. Seah, I. Welch, Reliability in wireless sensor networks: a survey and challenges ahead, Comput. Netw. 79 (2015) 166–187, https://doi.org/10.1016/j.comnet.2014.12.016.

[8] M. Katiyar, H.P. Sinha, D. Gupta, On reliability modeling in wireless sensor networks: a review, Int. J. Comput. Sci. Issues 3 (9) (2012) 99–105.

[9] A. Damaso, N. Rosa, P. Maciel, Reliability of wireless sensor networks, Sensors 9 (14) (2014) 15760–15785, https://doi.org/10.3390/s140915760.

[10] L.D. Xing, An efficient binary-decision-diagram-based approach for network reliability and sensitivity analysis, IEEE Trans. Syst. Man Cybern. Part A Syst. Hum. 1 (38) (2008) 105–115.

[11] L. A. Laranjeira and G. N. Rodrigues, Border effect analysis for reliability assurance and continuous connectivity of wireless sensor networks in the presence of sensor failures, IEEE Trans. Wirel. Commun. (13) 8 (2014) 4232–4246, doi:https://doi.org/10.1109/TWC.2014.2314102.

[12] C.N. Wang, L.D. Xing, V.M. Vokkarane, Y.L. Sun, A phased-mission framework for communication reliability in WSN, in: Proceedings of the 60th Annual Reliability and Maintainability Symposium (RAMS'14), 2014, pp. 1–7.

[13] J. Tripathi, C.O. Jaudelice, J.P. Vasseur, Proactive versus reactive routing in low power and lossy networks: performance analysis and scalability improvements, Ad Hoc Netw. 23 (2014) 121–144.

[14] M. Yan, K.-Y. Lam, S. Han, et al., Hypergraph-based data link layer scheduling for reliable packet delivery in wireless sensing and control networks with end-to-end delay constraints, Inf. Sci. 278 (2014) 34–55.

[15] P. Schaffera, K. Farkas, Á. Horváth, T. Holczer, L. Buttyán, Secure and reliable clustering in wireless sensor networks: a critical survey, Comput. Netw. 56 (11) (2012) 2726–2741.

[16] S. Mahadevan, R. Rebba, Validation of reliability computational models using Bayes networks, Reliab. Eng. Syst. Saf. 87 (2) (2005) 223–232.

[17] S. Distefano, Evaluating reliability of WSN with sleep/wake-up interfering nodes, Int. J. Syst. Sci. 44 (10) (2013) 1793–1806, https://doi.org/10.1080/00207721.2012.670293.

[18] J.L. Cook, J.E. Ramirez-Marquez, Reliability analysis of cluster-based ad-hoc networks. Reliab. Eng. Syst. Saf. 93 (10) (2008) 1512–1522, https://doi.org/10.1016/j.ress.2007.09.002.

[19] Y. Song, T. Chen, M. Juanli, Y. Feng, X. Zhang, Design and analysis for reliability of wireless sensor network, J. Networks 7 (12) (2012) 2003–2012.

[20] I. Kabashkin, J. Kundler, Reliability of sensor nodes in wireless sensor networks of cyber physical systems, Procedia Comput. Sci. 104 (2017) 380–384, https://doi.org/10.1016/j.procs.2017.01.149.

[21] C.F. Chiasserini, I. Chlamtac, P. Monti, A. Nucci, Energy efficient design of wireless ad hoc networks, in: Proceedings of the 2nd International IFIP-TC6 Networking Conference on Networking Technologies, Services and Protocols: Performance of Computer and Communication Networks: Mobile and Wireless Communications, May 19–24, 2002, Pisa, Italy, 2002, pp. 376–386.

[22] M. Slovick, Buck-Boost Controller Answers Call for Redundant Battery Systems, Electronic Design, https://www.electronicdesign.com/automotive/buck-boost-controller-answers-call-redundant-battery-systems, 2018. (accessed 29.09.19).

[23] S. Taranovich, Redundancy Design Solution for Battery Systems in Autonomous Vehicles, https://www.edn.com/electronics-products/electronic-product-reviews/other/4461118/Redundancy-design-solution-for-battery-systems-in-autonomous-vehicles, 2018. (accessed 29.09.19).

[24] A. Mahapatro, P.M. Khilar, Fault diagnosis in wireless sensor networks: a survey, IEEE Commun. Surv. Tutorials 15 (4) (2013) 2000–2013.

[25] I. Kabashkin, Reliability of cluster-based nodes in wireless sensor networks of cyber physical systems, Procedia Comput. Sci. 151 (2019) 313–320. Elsevier, https://doi.org/10.1016/j.procs.2019.04.044.

[26] R. Barlow, K. Heidtmann, On the reliability computation of a k-out-of-n system, Microelectron. Reliab. 33 (2) (1993) 267–269.

[27] K. Misra, Handbook of Performability Engineering. Springer, 2008. https://doi.org/10.1007/978-1-84800-131-2.

[28] P. McGrady, The availability of a k-out-of-n:G network, IEEE Trans. Reliab. 5 (R-34) (1985) 451–452.

[29] A. Rushdi, A switching-algebraic analysis of consecutive-k-out-of-n:F systems, Microelectron. Reliab. 27 (1) (1987) 171–174.

[30] M. Ayers, Telecommunications System Reliability Engineering, Theory, and Practice, Wiley-IEEE Press, 2012.

[31] B. Kozlov, I. Ushakov, Reliability Handbook (International Series in Decision Processes), Holt, Rinehart & Winston of Canada Ltd., 1970.

[32] M. Farsi, M. Badawy, M. Moustafa, H. Arafat Ali, Y. Abdulazeem, A congestion-aware clustering and routing (CCR) protocol for mitigating congestion in WSN, IEEE Access 7 (2019) 105402–105419, https://doi.org/10.1109/ACCESS.2019.2932951.

[33] B. Bhushan, G. Sahoo, Recent advances in attacks, technical challenges, vulnerabilities and their countermeasures in wireless sensor networks, Wirel. Pers. Commun. 2 (2017) https://doi.org/10.1007/s11277-017-4962-0.

[34] P. Neamatollahi, S. Abrishami, M. Naghibzadeh, M.H. Yaghmaee Moghaddam, O. Younis, Hierarchical clustering-task scheduling policy in cluster-based wireless sensor networks, IEEE Trans. Ind. Inf. 14 (5) (2015) 1876–1886, https://doi.org/10.1109/TII.2017.2757606.

[35] C. Tunca, S. Isik, M.Y. Donmez, C. Ersoy, Ring routing: an energy-efficient routing protocol for Wireless Sensor Networks with a mobile sink, IEEE Trans. Mob. Comput. 14 (9) (2015) 1947–1960, https://doi.org/10.1109/tmc.2014.2366776.

[36] B. Bhushan, G. Sahoo, An acknowledgement-based mobile sink routing protocol with rechargeable sensors for wireless sensor networks, J. Mobile Commun. Comput. Inform. 5 (2019), https://doi.org/10.1007/s11276-019-01988-7.

[37] K. Singh, K. Singh, L.H. Son, A. Aziz, Congestion control in wireless sensor networks by hybrid multi-objective optimization algorithm. Comput. Netw. 138 (2018) 90–107, https://doi.org/10.1007/s12083-019-00758-8.

[38] Z. Zhang, A. Mehmood, L. Shu, Z. Huo, Y. Zhang, M. Mukherjee, A survey on fault diagnosis in wireless sensor networks, IEEE Access 6 (2018) 11349–11364, https://doi.org/10.1109/ACCESS.2018.2794519.

[39] G. Rubino, B. Sericola, Markov Chains and Dependability Theory, Cambridge University Press, 2014. https://doi.org/10.1017/CBO9781139051705.

Lightweight cryptographic algorithms for resource-constrained IoT devices and sensor networks

8

Pulkit Singh[a], Bibhudendra Acharya[a], and Rahul Kumar Chaurasiya[b]

[a]*Department of Electronics and Communication Engineering, National Institute of Technology, Raipur, Chhattisgarh, India* [b]*Department of Electronics and Communication Engineering, Malaviya National Institute of Technology, Jaipur, Rajasthan, India*

Chapter outline

153

Security and privacy issues in IoT devices and sensor networks. https://doi.org/10.1016/B978-0-12-821255-4.00008-0

1 Introduction

Internet of Things (IoT) has been a model of typical entities capable of detecting and interacting Internet-based smart appliances. It combines information gathered from several gadgets and utilizes statistics to exchange information with databases [1]. The IoT carries the cellular network, chips, sensor network, and conventional Internet to a modern stage as it connects everything to the Internet [2]. It has various applications in human life and in industries. One area of concern that needs to be addressed is to guarantee the confidentiality, data integrity, and authenticity issues that will arise as a result of security and privacy [3].

In the recent past the Internet has become more and more popular among smart devices. The world has connected from everywhere using Internet, even houses are filling with smart locks, smart TVs, smart phones, computers, etc. In future, these appliances would be monitored through Internet. Home appliances such as air conditioners and refrigerators will also be controllable over the Internet [4]. It can be assumed that billions of objects will be connected to the Internet in future [5]. This implies that there will be many things online for each individual on earth. The world will be blanketed with millions of sensors for collecting and uploading data from physical world to the Internet. Currently, the IoT implementation is still in the primary stage but is quickly evolving with time [6, 7]. IoT helps automobile industries to develop modern automation techniques [8]. A special application of IoT has been in the healthcare sector and there will be enormous challenges for enhancement in healthcare [9–11]. It can be expected that IoT will contribute for constructing mining operations safer and it will be feasible to predict the occurrence of disaster as well [12]. Automotive services and transport technologies are planned to be developed with IoT [13]. To monitor and track the motion of objects from origin to destination, physical devices will also be mounted with sensors and RFID tags for enabling transport businesses [14]. The IoT demonstrates promising behavior in the logistics industry. Hence, there are many users eager to adopt IoT applications with the intention of contributing in economic growth, healthcare facilities, transportation, and a better lifestyle. IoT must provide sufficient security to motivate the process of adaptation for all smart applications.

In addition, a sensor network contains group of tiny nodes responded via wireless medium depending on programming. These nodes have less physical space, low processing power, and small storage capability. Due to this, sensor network requires limited amount of computational power, storage, energy, and communication bandwidth. Therefore lightweight and energy-efficient cryptographic algorithms are a good choice. Moreover, wireless multimedia sensor network (WMSN) is useful for healthcare monitoring systems. This network uses low-cost cameras for transmitting multimedia audio and video streaming as well as high-resolution images. Therefore multimedia network makes the balance between security, energy cost, and application purposes. Hence the new branch of cryptography, namely, lightweight cryptography (LWC) forces to make trade-off between cost, security, and performance. On the other hand, one class of sensor network uses sensor nodes for surveillance purposes and real-time applications, known as wireless visual sensor network (WVSN). This network is highly concerned to process image signals having low latency of communication. To secure resource-constrained real-time systems of this sensor network requires high-speed LWC.

Over the years, modern technology has developed from one stage to next stage in IoT evolution such as from handwritten letters to telephone operators, manually linking calls to mobile phones, cellular phones to smart phones, etc. [15]. On the other hand the chances of attack on the users' privacy have also been increased with the introduction of fresh technologies in IoT and in sensor network. Moreover the need for protection and reliable techniques in communication sector is now more essential [16, 17]. Therefore, cryptography plays a significant role in making secure communication. It ensures end-to-end user by providing high security of their confidential data and sensible information. Basically, cryptography is about encryption and decryption. Encryption is a procedure of transforming a one form of message into another form. Decryption is reverse operation of encryption process. Importantly, cryptography is categorized into two categories: symmetric key cryptography, also known as private key cryptography, and asymmetric key cryptography, also known as public-key cryptography [18].

A conventional cryptographic algorithm, known as advanced encryption standard (AES), provides high security in network servers, desktops, tablets, and smart phones, etc. [19]. These devices process and keep large data requiring reasonable processing power, large memory space, and physical area. Moreover, these devices also transform critical and sensible information. A weak cryptographic algorithm is not capable to provide security for these devices, and adversary can breach security. In addition, conventional cryptographic algorithms require more computational power and have high latency. However, IoT devices contain embedded devices that have restricted resources in terms of energy, memory, and processing power. Hence a new branch of cryptography has been developed, which enables algorithms for resource-constrained applications. These applications are likely to force small amount of information to be encrypted. In this resource-constrained environment, cryptographic algorithms should have low computational complexity. Due to lack of critical information, moderate security is sufficient for such applications [20].

1.1 Methodology

The major contribution of this chapter is to discuss utilization of different design strategies by developing three architectures for KLEIN lightweight block cipher. These architectures enhance the performances in terms of speed, area, and energy. The proposed pipelined architecture is designed by putting finite number of registers into algorithm. This idea reduces critical path delay between two operations. Therefore, pipelined architecture design is capable of providing high operating frequency. Next, parallel architecture implementation reduces cost of the design by achieving better result in terms of throughput per slice metric. This design decreases the number of switching activities due to lesser computations. Additionally, a loop unrolling architecture is a technique of loop transformation that helps to improve the execution time of a system. The proposed loop unrolled design strategy is beneficial for energy efficient implementation. This implementation results in a better architecture with reduction in energy per bit metric. Unrolled architecture implementation enables unrolling of multiple rounds, up to total number of rounds required by the cipher.

1.2 Chapter organization

The literature survey of various lightweight block ciphers used for resource-constrained applications (like IoT) is discussed in Section 2. Section 3 presents various phases and layers of IoT, where lightweight algorithms may be used for providing adequate security. Section 4 summarizes the primitives of lightweight block ciphers with their different designing blocks. The three proposed design architectures of KLEIN lightweight block cipher are explained in Section 5. This section also describes the simulation environments and depicts the results of the proposed designs in various hardware metrics. Section 6 compares the performance of conventional and lightweight block ciphers for resource-constrained applications. Finally, Section 7 concludes the chapter and provides the future directions.

2 Related work

IoT technology has been included in several sectors such as logistics, smart city, industrial manufacturing, smart healthcare, automation, transportation, e-education, e-governance, retail, and business management. Infect, IoT has contributed to the development of each sector connected to the Internet [21, 22]. In addition, IoT system generally involves the interaction of huge distinct system mechanisms and technologies in various conversation contexts (viz., person to things, thing to things, or person to persons) [23]. Moreover the recent literatures have defined IoT as a region in which physical objects are progressively incorporated to create data systems. Authors have specified the purpose to support innovative and smart facilities to customers [24]. In this way the interconnected *things* collect, observe, and analyze all forms of surrounding information with the help of variety of nodes like sensors and actuators. The information is collected in real time from objects, people, animals, and plants. This information needs to be secure in communication network. Hence, lightweight cryptographic algorithm plays an essential role to secure this type of network.

Before introducing lightweight ciphers for the IoT applications, the conventional ciphers such as AES and DES have been employed in IoT [25]. In [26, 27], authors have made fast software implementations for 8-bit and 16-bit microcontrollers and an application-specific integrated circuit (ASIC) implementations for AES-128 cipher. However, the requirement of constrained resources in IoT applications makes the conventional ciphers unfit in the development of such technology. Hence the employment of lightweight cryptographic algorithms were required to protect these applications [3]. In this context the IoT limitations have been discussed in literature [28, 29], and focus has been drawn on the requirements of the lightweight cryptographic algorithms. In essence, there are some lightweight cryptographic algorithms that do not always take advantage of trade-offs in security and efficiency. However, the lightweight block cipher has shown substantially better appearances over lightweight hash functions and stream cipher [30].

Many lightweight block ciphers have been developed for securing resource-constrained IoT applications. An existing literature has described the utility of lightweight block cipher on 64-bit processors. Authors have demonstrated that the hardware- and software-based devices communicate well with appropriate security for a back-end server [31]. Some lightweight block ciphers, namely, SIMON and SPECK, can be used in flexible mode and are suitable for IoT applications [32]. There are several lightweight block ciphers that have set benchmark for IoT applications, namely, HIGHT [33], LBlock [34], MIBS [35], Piccolo [36], QTL [37], RECTANGLE [38], PRINT [39], PRINCE [40], mCrypton [41], Twine [42], PRESENT [43], SFN [44], LILLIPUT [45], KLEIN [46], Midori [47], and LEA [48]. Each of these ciphers works with comprehensive approaches, and also, these block ciphers have unique features or properties for resource-constrained applications in IoT [49].

3 Preliminaries

3.1 Resource constrained environment- internet of things (IoT)

The application of IoT has created resource-constrained environment in different sectors. This environment includes all types of processing devices, starting from a sensor to the latest cloud technology. These processing devices have completely changed the technologies of modern human life. The applications of these devices can be seen in various aspects of modern world including smarts cards and routine cards, vehicles parking and tracking, taking care of children and elders, and patient care (from entry till postsurgery). Additionally, these devices have also influenced the humans' feeling. IoT applications cover a wide variety of systems, namely, vehicular, ubiquitous, grid, and distributed.

The sensors, RFID tags, nanotechnologies, and smart technologies are the key features of the developments of IoT for a variety of services. Moreover, IoT technology has become easily accessible due to significant decrement in the cost of different devices, chips, sensors, and smart phones. Several other factors like the requirement of low processing power and smaller bandwidth, migration toward 5G and Internet Protocol version-6 (IPv6) address space are also suitable for IoT applications. The devices and technologies used for IoT applications are different for each phase and layer as shown in Fig. 1. IoT incorporates "physical and virtual" world anywhere and anytime. Therefore, it introduces "maker and hacker" for attention. Since, IoT

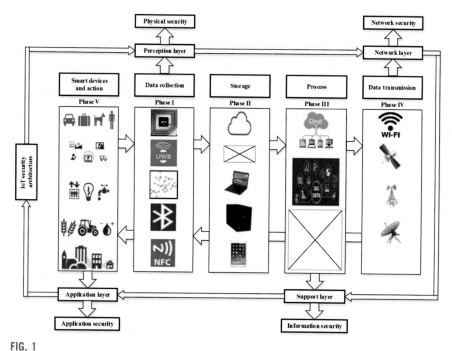

FIG. 1

IoT enabling technology: phases and layers.

observes all types of machines and human beings in similar way, the security is a serious problem when two objects are combined to the Internet.

3.1.1 Phases of IoT system

From the data collection to delivery, IoT system can be divided into five phases as shown in Fig. 1. During these phases the network uses different devices and protocols to ensure the security and efficient communication. This section deals with the detailed description of each phase.

- *Phase I:* Data acquisition and perception
  This phase is responsible for the collection of data from different devices. To acquire data the devices should be capable of transmitting information into network. For acquisition the static devices generally use RFID tags and body sensor, whereas dynamic devices use sensors and chips. The different enabling technologies, namely, WSN, Bluetooth, near-field communication (NFC), and ultrawide band (UWB) are refereed in this phase.
- *Phase II:* Storage
  In this phase the collected data are stored in various storing elements. The IoT network uses in-built memory space that mostly has limited processing

capabilities and small storage capacity. The IoT system also stores the data into cloud for further analysis.

- *Phase III:* Intelligent processing
  The stored data are analyzed and converted into a control signal for intelligent services for the user. Hence, this phase responds the queries in real time and sets the management protocol for data processing. This phase plays an important role in discriminating between physical and the real world.
- *Phase IV:* Data transmission
  In this phase the transmission of stored, analyzed, and processed data takes place. In other words, that data are transformed in the IoT network between different devices and across the phases. Different technologies like Ethernet, Wi-Fi, cellular network, satellite network, and vehicular ad hoc network (VANET) are employed for data transmission.
- *Phase V:* Smart devices and action
  The processed data transmitted over the network should be responded to a user connected through the Internet. This application generally involves the interaction of huge distinct system mechanisms and technologies in various conversation contexts. This includes person-to-things, thing-to-things, things-to-person, or person-to-persons interactions. This phase allows taking action based on the data collected from people, things, animals, and plants in real time.

3.1.2 Architectural layers of IoT

As depicted in Fig. 1, the IoT architecture enables their services by dividing the functionality into four layers, namely, perception, network, support, and application layers [50]. These layers are described in detail in this section.

- *Physical/perception layer:*
  The function of this layer is to gather information from physical objects. The key features are to sense the physical equipment by capturing information and represent the physical world in the digital world. To enable these devices through Internet, different techniques are used to scan the stored data. To collect the data, physical devices such as sensors, Bluetooth, RFID tags, UWB, GPS, and NFC are used. Essentially the physical layer not only offers the data transmission services but also handles the management mechanisms. This holds the database of physical devices and gives permission to other layers' management entity. Moreover, IEEE 802.15.4 standard is utilized as an IoT specification in this layer. This standard provides adequate security, namely, physical security, but still demands to resolve some of its current attacks.
- *Network layer:*
  The network layer is the second layer in the IoT architecture, which collects information from physical layer via Internet. The collected data are transformed through Ethernet, Wi-Fi, satellite network, cellular network, VANET, cognitive radio network (CRN), etc. The security and transmission protocols are required by enabling Datagram Transport Layer Security (DTLS) for end-to-end

communication. Generally, IoT technology uses IPv4 address space over IPv6 addressing instrument. This layer divides the message into bundles and routes the packets from source to destination. Cryptographic algorithms protect the data transmission by utilizing Internet Protocol security (IPsec) at this layer. The network layer is highly prone to attacks, that is, related in routing protocol, and address compromise. In addition, bandwidth attacks such as distributed denial-of-service (DDoS) and denial-of-service attack (DoS) can initiate serious traffic crowding and also affect regular activities, leading to disruption in communication [51].

*   *Support layer:*
    This layer provides a supporting system for the network layer and application layer. Almost all type of cryptographic algorithms and antivirus systems are situated in this layer to securely transfer the data to application layer. Additionally, mass data processing and intelligent decisions are incorporated in this layer for different computing systems such as cloud computing, intelligent computing, big data. This layer should be strong enough for recognizing malicious activity to ensure that the intelligent processing remains unaffected.
*   *Application layer:*
    This is the top-most layer in IoT architecture and offers facilities to the users. The users can use IoT applications by connecting devices through this layer from their devices such as mobile phone and laptop. This layer decides the levels of security according to the accessing requirements of different applications. High security is provided to the applications requiring more access. In addition, amount of data sharing makes a choice for intruder in this layer, which can introduce the issue in access control, data privacy, and leakage of intelligent data.

3.2 Implementation of lightweight block ciphers for IoT applications

IoT has filled the gap between the physical world and the virtual world that interact all physical objects such as smart TVs, smart doors, cars, tablets, and people, connected with Internet. IoT defines these objects in the virtual world by taking their features and connects all from anywhere in the world at any time. Due to fewer requirements of hardware and power consumption with affordable security for this technology, different lightweight cryptographic algorithms are suitable for such developments. The symmetric lightweight block ciphers use simple primitives like shifting, rotation, permutation, substitution, and bit-wise XOR.

The best suggestion of utilizing symmetric block ciphers for resource-constrained IoT devices and sensor networks is to use light eight cryptography (LWC). There are two main reasons to support this proposal: applicability for lower resource devices and end-to-end communication efficiency.

*   *Applicability for lower resource devices*
    The lightweight cryptographic primitives have a lower footprint than standard cryptographic primitives. The network links have the ability to process lower

footprint only. The lightweight cryptographic primitives can offer opportunities with reduced resources for other network links.

- *End-to-end communication efficiency*
  IoT is used for interaction of end-to-end services. The communication between end nodes is more efficient by introducing a symmetric key algorithm in case of LWC. In addition, the cryptographic approach is essential for low-resource systems with a restricted amount of power requirement in battery-powered devices. Therefore the introduction of the symmetric key encryption scheme in this branch of cryptography enables the reduction in power requirement for end-node devices.

In addition, wide range of processors can communicate information securely between end-nodes efficiently for software applications. Such instances can be considered in WSN systems. However, ASICs demand lowest cost implementation, where hardware characteristics are crucial due to limited cost and low power consumption. Based on target applications, lightweight characteristics are defined in ISO/IEC 29192 standard [52]. The significant steps for evaluating the lightweight characteristics are chip size and/or power usage in hardware solutions. The smaller code and/or random-access memory (RAM) storage is acceptable for lightweight devices in software solutions. The lightweight primitives for Internet security guidelines like DTLS and IPsec are better than standard cryptographic primitives from the perspective of the implementation characteristics. On the other hand, reasonable safety is provided by lightweight cryptographic algorithm. There is still a gap to develop efficient lightweight cryptographic primitives. This is due to a necessity that the LWC will provide an optimum balance in the trade-off of security and efficiency [28].

Various lightweight block ciphers can be used for implementing IoT applications. Almost all of the blocks ciphers convert 64 bits of data into an intelligent form using either 80-bit or 128-bit key sizes. It is clear that there is limited number of rounds in lightweight ciphers for reducing the computational complexity as shown in Table 1. In addition, block ciphers have different type of structures in their round function such as generalized Feistel network (GFN), Feistel structure, substitution-permutation network (SPN), hybrid structure, and extended generalized Feistel network (EGFN). In these structures, GFN and EGFN structures are the advanced variants of Feistel structure and make better diffusion of bits [53]. SPN structure is better than Feistel structure in terms of hardware cost by using almost same hardware for encryption and decryption algorithms. On the other hand, Feistel structure performs well concerning security by introducing inner auxiliary function in the round. The lightweight ciphers can also be widely classified into two categories based on availabilities and applications in IoT, that is, stream ciphers and block ciphers. Stream cipher needs twice the internal states for achieving the same security as a block cipher. It is conceivable to realize relatively compact and exceptionally nonlinear functions in stream lightweight algorithms, while block ciphers are most likely to be used to implement substitution-box (S-box) more efficiently in hardware.

Table 1 Summary of some lightweight block ciphers.

Lightweight algorithms	Structure of round function	Block length (in bits)	Key length (in bits)	Number of rounds
HIGHT	GFS	64	128	32
LBlock	Feistel	64	80	32
MIBS	Feistel	64	64/80	32
Piccolo	GFN	64	80/128	25/31
QTL	SPN	64	64/128	16/20
RECTANGLE	SPN	64	80/128	25
PRINT	SPN	48/96	80/160	48/96
PRINCE	SPN	64	128	12
mCrypton	SPN	64	64/96/128	12
Twine	Feistel	64	80/128	32
PRESENT	SPN	64	80/128	31
SFN	Hybrid	64	96	32
LILLIPUT	EGFN	64	80	30
KLEIN	SPN	64	64/80/96	12/16/20
Midori	SPN	64/128	128	16/20
LEA	Feistel	128	128/192/256	24/28/32

S-box is only nonlinear function enhancing security in cryptographic algorithms. Therefore block cipher is highly concerned in providing more security and having efficient hardware implementation in IoT applications [18].

4 Lightweight cryptographic primitives

LWC has been a very important field to explore and focus on the scheme of designing new ciphers. This domain fits into the criteria given by the constrained environments. The terms like low computational capability, smaller area, low power consumption, and smaller code size are the key requirements defined in the use of constrained devices. Therefore the concept of lightweight stands for these basic needs rising from conventional cryptographic algorithms. Under these resource constraints, there is also a rising requirement for security problem described in lightweight algorithms that are defined depending on constrained requirements. Therefore this area highlights the successful deployment of cryptographic algorithms. It is a considerably recent research subfield combining the cryptography, computer science, and electronic engineering fields. The designers and research commodity of lightweight cryptographic field deal with making the balance between cost, security, and performance. Practically, two of the three design goals can probably be effectively improved, but at the same moment, enhancing all three-design goals is a challenging task for this field.

A designer of a lightweight system tries to balance the trade-off between cost, security, and speed as shown in Fig. 2. The cost mainly defined by data size and implementation strategies. Smaller key, lesser data size, and serial implementation make the system less expensive. Level of security depends on the key size and algorithmic complexity. The large key size and more number of rounds provide high privacy and security. Performance is directly related to speed. Parallel architectures and reduced number of rounds speed up cryptographic algorithms. However, the parallel architecture increases the cost and reduction in the number of rounds decreases the security. In other words an attempt to reduce the system costly may result in slow algorithm with higher chances for vulnerability. However, lightweight designers target to provide moderate security in constrained devices. Hence, several researches are being carried out for reducing the hardware area and power consumption, while providing a cost efficient algorithm with sufficient security. Lightweight cryptography field is further categorized based on security services like data integrity, authentication, and confidentiality.

The lightweight cryptographic primitives developed in recent past provide better performance over conventional ciphers for resource-constrained applications. As depicted in Fig. 3, some of these primitives are stream cipher, hash function, block cipher, and message authentication code. Lightweight primitives are dissimilar from standard algorithms. In summary the objective of lightweight weight cryptography is that designer can make optimum balance between cost, security, and performance for resource-constrained environments.

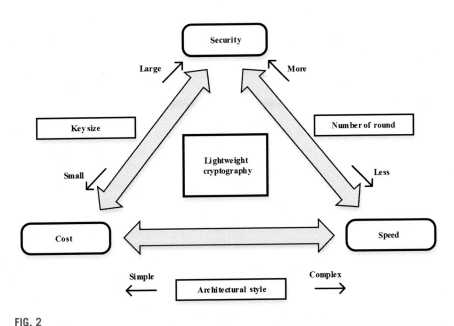

FIG. 2

Lightweight cryptography trade-off.

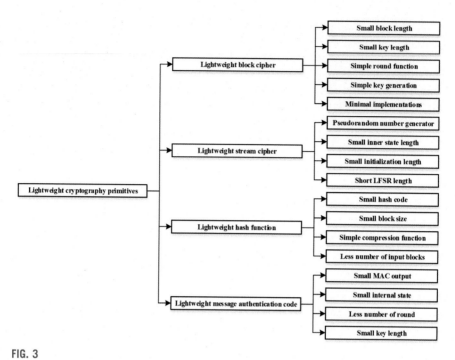

FIG. 3

Lightweight cryptographic primitives.

4.1 Lightweight block ciphers

The development of algorithms for lightweight block ciphers is motivated by the requirement of security for constrained applications. The conventional cryptographic block cipher such as AES provides high security but requires large area footprint. Hence, several lightweight algorithms have been developed fulfilling need over conventional ciphers. Among them, two-block ciphers PRESENT and CLEFIA have introduced in ISO/IEC 29192 standards for LWC. Some lightweight ciphers are designed removing the complexity of conventional ciphers [52]. For example, DESL is a revised version of DES, where the initial and final permutations have been removed and only one S-box is used in each round of the cipher. This results in designing lightweight hardware [54]. Some of the lightweight block ciphers, namely, RECTANGLE, mCrypton, LBlock, and TWINE are designed for achieving better performance in both software and hardware aspects. Some ciphers like SIMON and SPECK [55] and KATAN and KTANTAN [56] are developed to be flexible and simple. On the other hand, some lightweight block ciphers are targeted to achieve design goals for constrained hardware environments. PRESENT, QTL, and SFN are some of the examples for these ciphers.

There are different design options that make lightweight block ciphers preferable over conventional cryptographic algorithm for resource-constrained environment. The design options are described in the succeeding in detail.

- *Small block length:*
  The input message length of lightweight cipher is decided by the information to be encrypted. The standard block length for lightweight ciphers is kept to be 64 bits instead of 128 bits and 256 bits. The small block size saves the requirement of storing elements during the execution of cipher. Small block size may exhibit some attacks successfully such as known-plaintext, chosen-plaintext, and key recovery attacks.
- Small key length:
  Lightweight block ciphers are designed for providing moderate security in resource-constrained devices. Therefore 80-bit key length is considered to be sufficiently enough to provide adequate security. The relatively small key length saves the memory requirement at intermediate stages in cipher algorithms. However, if the key length is further reduced, then it would be prone for attacks like brute-force attack, related-key attack, and algebraic attack.
- *Simple round function:*
  The contents of the round function in LWC are very similar to conventional cryptographic block ciphers. The differences lie in the number of input-output bits of S-boxes and selection of permutation layer. Generally, 4-bit S-box is preferred in LWC for simplification at hardware side. Moreover a bit-wise permutation and recursive MDS matrix is preferred over the complex permutations. To avoid the compromise in security, the number of iterations may be increased for a simple round function.
- *Simple key generation:*
  Security of a cryptographic algorithm is proportional to the complexity of key scheduling algorithms. However, more complex key scheduling requires additional area, power, and computation time. Therefore, almost all the lightweight block ciphers use key algorithm generated by on-the-fly mode or direct mode. By this approach, design area and power consumption is reduced with no compromise in speed.
- *Minimal implementations:*
  Some lightweight block ciphers reuse basic building blocks of encryption round function in decryption process. This methodology helps to reduce implementation cost.

4.2 Lightweight stream ciphers

A stream cipher integrates binary input message with digital keystream generated by the pseudorandom generator. It transforms input message stream one bit or one byte at a time. Various lightweight stream ciphers have been developed for constrained environments promising primitives. The *eSTREAM* project started by the EUE-CRYPT network developed new stream ciphers fitted for worldwide selection. Under this project, stream ciphers such as Grain, MICKEY, and Trivium were developed for hardware platform with limited power consumption, area, and memory requirements [57].

There are important designing characteristics like autocorrelation, linear complexity, and unknown period that decide the security of stream ciphers. The necessary design factors to achieve the design of lightweight stream cipher are described in the succeeding text.

- *Pseudorandom number generator:*
  A pseudorandom number generator provides significant contributions for establishing an important role in encryption for different security requirements. This utilizes the feature to generate a deterministic bits stream that repeats after large period. The larger the repetition period, the harder it is to do cryptanalysis. Most of the lightweight stream cipher makes use of linear-feedback shift register (LFSR) and nonlinear-feedback shift register (NFSR) for producing stream of bits due to their simple operation and implantation.
- *Small inner state length:*
  The length of the internal state decides the need for additional space in limited resource applications. To accomplish it, the design of a new stream cipher was brought with the shorter internal condition on Fast Software Encryption (FSE) 2015 event [58]. On the other hand, Sprout stream cipher has been found to be insecure, designing with smaller internal size [59]. Therefore a new idea is introduced making cipher more secure. The idea is to utilize the key not only in the initialization phase but also in the generation of the keystream. However, the minimal inner state brings the weakness in the stream cipher. The requirement of making stream cipher more secure against time-memory-data trade-off (TMDTO) attack is that the length of inner state should be at least twice its level of security [60].
- *Small initialization length:*
  The size of the initialization vector also settles the resource utilization. For example, in LIZARD stream cipher, initialization length IV is smaller than the key length, achieving fewer hardware applications, whereas size of IVs larger than 64 bits does not need even in the context of packet mode. Moreover, Grain v1 cipher also operates on 64-bit initialization length IVs making better against hardware cost. However, for instances where it is necessary to encrypt more than 264 packets, this allows the use of large session keys and can increase the implementation cost. Maximum 2^{18} packets are sufficient for resource constrained applications [61].
- *Shorter LFSR length:*
  The period of keystream generator and linear complexity of stream cipher are decided by the size of the LFSR in bits. Due to the limitation of packet mode, the small length of LFSR in LIZARD cipher does not provide large periods as the

Grain family. Moreover, a large number of periods means more secure against attacks such as related-key attack, cube attack, classical TMDTO attack, and algebraic attack, [61], but it results in additional area overhead and computation complexity.

4.3 Lightweight hash functions

A hash function processes a message of variable length as input and converts it into an output of a specific length known as a hash code or message digest. It uses no key, but it is a function of initialization vector IV and message. In [62], authors have developed a hash function in RFID protocols without using the lightweight scheme. Since a standard hash function consumes more power and requires large processing state size, it is not suitable for limited-resource devices. Consequently, authors have designed a lightweight hash function for these resource-limited devices [63]. Some of the lightweight hash functions are Lesamnta-LW [64], SPONGENT [65], PHO-TON [66], and Quark [67].

A hash code algorithm is a subset of all message bits and gives a specified size of output bits. There are different design options that make lightweight hash function better in performance over conventional hash algorithms such as:

- *Small hash code:*
  The balanced size of hash code can be utilized for some applications, where collision resistance is not important. It also reduces the size of the internal state and saves the hardware cost. On the other hand the large size of hash code provides more security for applications that require collision resistance. A 88 bits of hash code of SPONGENT family of lightweight hash function is only secure against preimage resistance with best hardware performance [68]. The conventional cryptographic hash function like Secure Hash Algorithm-512 (SHA-512) delivers large size of hash code [69]. This size gives better guarantee by increasing computational complexity against brute-force attack [70].
- *Small block size:*
  A serialized architecture of hash function provides very compact implementations. For example, the serialized SPONGENT hash function is smaller in hardware performance than the QUARK hash function. To get lightweight hash function, the standard size of the data should be smaller, but this can reenforce the poor performance and even reduction in size of hash code as well. Therefore the chances of cryptanalysis increase by reducing the complexity against attacks.
- *Simple compression function:*
  The compression function combines the output of previous function with next input message block and then processes it for a finite number of rounds. However, in this approach, the complexity and large number of rounds require additional area. It also results in increment in power consumption and computational time.

Therefore a hash function algorithms can be used for resource-constrained environments by using the simple functionality of the compression algorithm.

- *Less number of input blocks:*
  The compression function processes the data iteratively with a finite number of round functions. A hash function divides input message into finite number of blocks with fixed sized. If the last block has lack of some bits from the size of each block, algorithm pads the required bits. The inclusion of extra bits into last block makes the job difficult for attacker. Less number of input blocks is feasible for lightweight hash algorithms.

4.4 Lightweight message authentication codes

A message authentication code (MAC) function is a symmetric primitive, which is used to ensure that messages are genuine or valid. It takes the message and a secret key as inputs and converts it into a short tag as output. Thus the fixed-length tag and key perform the integrity and authenticity of the message. Various authenticated encryption algorithms have been proposed for resource-limited applications such as RFID tags and wireless sensor nodes. Some algorithms provide authentication and encryption simultaneously, whereas some are only designed to be used for authentication purpose by generating and verifying the tag for each message processed [71]. The designing of MACs has utilized three approaches. Firstly, message authentication algorithm (MAA) and more recently universal message authentication code (UMAC) algorithms use new primitives in their designs. Such a technique makes it possible to optimize the trade-off between security and performance. In second approach, existing primitives employ either an unkeyed hash function or a new mode of operation, for example, NMAC, HMAC and XCBC, CBC-MAC, RMAC, and OMAC. Finally, New MACs like two-track MAC and MDx-MAC are designed using existing primitive elements [72].

Some design considerations make lightweight MAC function better in performance over conventional MAC algorithms. These design options are described as follows:

- *Small MAC output:*
  The output authentication tag length defines the balance in area and security. Theoretically, it is found that the length of the MAC output can be anything between 48 and 96 bits [72]. Larger MAC output values offer more guarantee against attacks, but its performance pays additional bandwidth/storage for the MAC. It is recommended that the size of MAC output should be at least 64 bits for reliable communication. However, as less than 64-bit output, MAC value can be used for resource-constrained IoT applications [73].
- *Small internal state:*
  The length of internal state decides the computational overhead and implementation cost in lightweight MAC algorithm. However, some MAC function like TuLP algorithm utilizes only 64-bit internal state [73]. It may not be

sufficiently large to protect against the birthday attack on internal states in conventional applications [74]. However, it is considered strong enough for some the wireless sensor applications. Still, to get assurance against attacks for other applications, TuLP-128 increases the authentication tag and internal state lengths at the cost of additional area requirement.

- *Small number of round:*
  The number of rounds affects the hardware cost in MACs. For example, in lightweight MAC algorithm TuLP, compression function is tuned to less number of rounds for balancing a trade-off between efficiency and security.
- *Smaller key length:*
  A complex key scheduling algorithm requires large hardware and consumes more power. The padding rules make key length dynamic. Therefore the smaller key length is also used for balancing the trade-off between performance security [73].

5 Proposed methodology

5.1 Hardware analysis metrics

The lightweight block ciphers for IoT applications are measured based on different metrics. These performance metrics can be broadly categorized into two design implementations, namely, hardware and software. Some of the metrics are useful in both the implementations, whereas others are helpful either in hardware or software implementation. In the case of IoT applications, design of lightweight block ciphers is mostly concentrated on hardware implementation. This is because the hardware implementation cannot be easily modified or read by an intruder. It offers higher physical security implementation. The different metrics evaluated in FPGA platform are described in the succeeding text.

- *Area:*
  Area is an important hardware implementation metric to evaluate a block ciphers. It is measured in number of slices, flip-flops, and LUTs. Some FPGA technology counts area in terms of number of logic elements. Area of a block cipher changes by changing the FPGA families and devices. The strategies used for circuit realization also affect the area of a block cipher. Several circuit realization techniques are employed in the implementation of ciphers. These implementations either utilize serial or parallel architectures. In general, area is useful parameter to define the cost of implementation.
- *Area per bit:*
  The number of input bits varies in some lightweight block cipher implementations. Therefore a block cipher may be assessed by area per bits (*area/bit*) parameter. Here, area is normalized with the number of bits at the input. Hence, this performance metric effectively assesses the cost of ciphers [75].

- *Power and Energy:*
  The average power is specified as the amount of energy transferred per unit time. In hardware implementation, power consumption increases with the size of a cipher. Minimizing the clock frequency decreases the power dissipation. However, it degrades the throughput [75]. On the other hand, energy consumption is an estimation of power consumption over total time. Energy can be computed in the encryption process by multiplying the power dissipation with the time required to produce the whole ciphertext and is expressed in micro Joule (μJ). Energy reflects a resource requirement, while power reflects a performance characteristic. For example, each encryption performed by a cipher requires some energy. A battery holds a finite amount of energy, can only support a limited number of encryption operations. Energy per encryption is used to predict the number of encryptions on a battery charge. Moreover, it is impossible to estimate battery capacity without computing power consumption [76].
- *Energy per bit:*
  This metric normalizes energy with respect to number of input bits in a block cipher [77]. Differentiating and preferable over energy metrics, energy per bit effectively demonstrates the total energy cost, measured in micro Joule per bit ($\mu J/bit$). Based on this metric an engineer can decide a basic primitive of block cipher that will complete the energy requirements computed by Eq. (1).

$$Energy\big/bit = Power \times Latency\big/OperatingFrequency \tag{1}$$

- *Throughput:*
  Throughput is defined as the number of output bits produced per unit time. It can be calculated in bits per second (*bps*). It is a measurement of the number of operations executed for getting a fixed result over a certain period of time. Throughput defines the speed of a cipher and is mathematically calculated as in Eq. (2).

$$Throughtput = Block\_size \times OperatingFrequency\big/Latency \tag{2}$$

The design that is able to produce high throughput is considered good in IoT application. The increment of clock frequency improves throughput.

- *Latency:*
  Latency is the time delay before getting the output after entering corresponding inputs of data into a cipher. Latency is calculated in the number of *clock cycles* required to encrypt the plaintext in cryptographic algorithms [77]. Less number of *clock cycles* results in low latency and enhances the speed of a block ciphers.

5.2 Algorithm of KLEIN lightweight block cipher

KLEIN is a lightweight block cipher [46] that performs the following three operations in a round function: *SubNibbles*, *RotateNibbles*, and *MixNibbles*. KLEIN cipher contains SPN and encrypts 64 bits of plaintext with three variants of keys of 64 bits, 80 bits, and 96 bits and 64 bits of ciphertext is produced at output after 12, 16, and 20 rounds, respectively. In the encryption process, plaintext of 64 bits is XORed with 64 bits of the key before performing round function in each round. For 80 and 96 bits of key scheduling, LSB 64 bits are extracted from the round keys. The output of XOR gate is divided into 16 nibbles that is later passed through S-boxes. This algorithm is processed based on byte orientation. During the *RotateNibbles* operation, two consecutive bytes are shifted left in each round. In *MixNibbles* step, similar to AES algorithm, mixing of columns is performed using matrix multiplication. Feistel-based structure is used for KLEIN key scheduling. Incremental round counter adds constants in each round. Subkeys are calculated by on-the-fly approach, which means that round transformation is performed at the time of generation of subkeys. The key scheduling of KLEIN for 64-bits is modeled as given in the succeeding text:

$$K = k_{63} \, k_{62} \, k_{61} \ldots k_2 \, k_1 \, k_0.$$
$$K_1 = k_{63} \, k_{62} \, k_{61} \ldots k_{34} \, k_{33} \, k_{32}.$$
$$K_2 = k_{31} \, k_{30} \, k_{29} \ldots k_2 \, k_1 \, k_0.$$
$$K_{11} = K_1 \lll 8 \ \& \ K_{21} = K_2 \lll 8.$$
$$K_{12} = K_{21} \ \& \ K_{22} = K_{11} \text{ xor } K_{21}.$$
$$K_f \, [44{:}40] = K_{12} \, [44{:}40] \text{ xor } RC^i.$$
$$K_f \, [11{:}08] = S \, (K_{22} \, [11{:}08]).$$
$$K_f \, [15{:}12] = S \, (K_{22} \, [15{:}12]).$$
$$K_f \, [19{:}16] = S \, (K_{22} \, [19{:}16]).$$
$$K_f \, [23{:}20] = S \, (K_{22} \, [23{:}20]).$$

In the earlier expressions, the f subscript in K_f denotes final key for the described rounds. RC^i is used for round constants representation with incremented value of i for different rounds. S denotes S-box input values.

Several hardware architectures of KLEIN block cipher are possible by using different design strategies. In addition, decryption process is reverse operation of encryption process. Using these design strategies, the architecture of KLEIN cipher can be improved in terms of speed, area, and energy. The proposed architectures are described in the following section.

5.2.1 Pipelined architecture of KLEIN lightweight block cipher

A pipelined architecture can be designed by having a finite number of registers in an algorithm. The architecture reduces the critical path delay between two operations. IoT processes many devices at same time and requires high throughput architecture. Pipelined design is capable of providing high throughput. As shown in Fig. 4 the pipeline registers are placed at the start of each round in KLEIN lightweight block cipher. The number of registers is varied in accordance with the number of rounds for

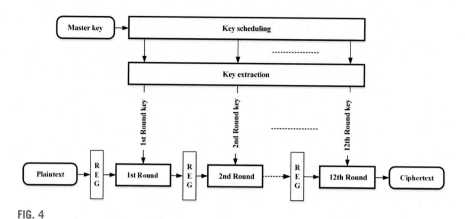

FIG. 4

Proposed Pipelined Architecture of KLEIN cipher.

different key scheduling of KLEIN cipher. Hence, 12 clock cycles are required to produce first ciphertext output for 64-bit key. The generated subkeys are available whenever round function adds the subkey. These registers are utilized for temporary storage of the data during the execution of rounds. Pipeline registers transfer both information and the control signal from one pipeline stage to the next. One stage of pipeline performs one operation of instruction; therefore each combinational operation is performed on behalf of instructions placed between any couples of pipeline registers. Consequently, pipelined registers transform data from one register to the next register in a single clock cycle. This process increases the operating frequency with high throughput and area requirement. To use a large number of pipelined registers, the hardware area is increased, which is computed in terms of number of flip-flops, LUTs, and slices. As the pipelining strategy reduces the critical path delay by expanding the possible frequency of the clock, it results in an allowable product of area and performance. Finally, it is observed that high operating frequency is achieved by selecting two main factors in the design. The first factor is to fix an ideal number of pipeline stages, and second is to find the best place to situate the pipelining register.

5.2.2 Parallel architecture of KLEIN lightweight block cipher

A parallel processing strategy decreases the cost by reducing the hardware area. This strategy also decreases the number of switching activities and required less number of clock cycles. Therefore hardware efficient architecture is achieved in terms of throughput per slice. Here, input of algorithm is the XOR of plaintext and input key. The resultant output is passed through combinational round function, which has following operations: S-boxes, *RotateNibbles*, and *MixNibbles*. The proposed architecture is depicted in Fig. 5. As in the figure, there are 16 S-boxes, which take 4-bit input and produce 4-bit output. The number of rounds in the parallel design depends on the length of a key. There are 12, 16, and 20 rounds for 64-, 80-, and

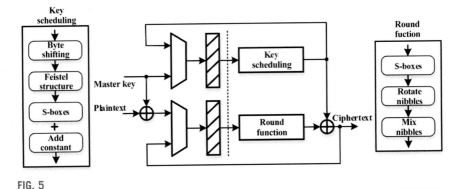

FIG. 5

Proposed parallel architecture of KLEIN cipher.

96-bit key, respectively. The output of a round function is stored in state register. Later the same output is applied to the same round function in next clock cycle. The length of output obtained from each operation and final ciphertext values are equal to size of plaintext. In addition, there is a key scheduling for KLEIN cipher, which means that subkey changes for each round. Key scheduling is also passed through different operation, namely, *ByteShifting*, *FeistelStructure*, and combination of *S-boxes* and *AddConstant*. The output of key scheduling is stored in another key register and is applied in key scheduling algorithm again for next clock cycle. Hence, this type of architecture generates ciphertext after 12, 16, and 20 clock cycles for 64-, 80-, and 96-bits key sizes, respectively. Since latency is equal to number of clock cycles, hence latency are 12, 16, and 20 for 64-, 80-, and 96-bit key sizes, respectively.

5.2.3 Unrolled architecture of KLEIN lightweight block cipher

An unrolling architecture enables unrolling of multiple rounds, up to total number of rounds required by the cipher. Hence a loop unrolling architecture is a method of loop transformation that helps to improve the execution time of a system. It removes or reduces iterations and increases the speed by eliminating loop control instruction. The proposed unrolled architecture of KLEIN cipher is shown in Fig. 6. This design is a technique that allows architecture to achieve large degrees of parallelism. A group of number of rounds can be realized within a single clock cycle without the need of a state register. Here, all round subkeys are available for each round function in same clock cycle. Therefore, due to decreased value of latency for unrolled implementation, low energy is required for processing a single block. Unrolled strategy is beneficial for energy-efficient design. This implementation helps to design an architecture that requires lesser energy per bit. Hence, unrolled design minimizes the required number of clock cycle for performing encryption process. However, it increases the hardware area (in terms of LUTs and slices) along with delay for the system.

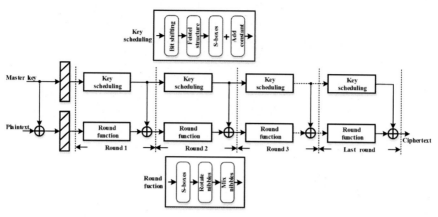

FIG. 6

Proposed unrolled architecture of KLEIN cipher.

5.3 Simulation results

The proposed architectures were synthesized in *Xilinx* using Integrated Synthesis Environment (ISE) design suite 14.6 and were simulated using *ISim* simulator. Virtex-4, Spartan-6, and Virtex-5 families were used with devicesxc4vlx25-12ff668, xc6slx16-3csg324 and xc5vlx50t-3ff1136, respectively, for comparison purpose. Each family has a certain number of slices, lookup tables (LUTs), and flip-flops configured in configurable logic blocks (CLBs). The architectures were evaluated after performing placement and routing. In addition, power consumptions were estimated using *Xilinx X Power* analyzer tool. The total power was computed as the sum of dynamic and static powers. Here the dynamic power includes the switching power and depends on the active current that flows during the switching process. The different architectures were applied in similar conditions for a fair comparison. These similar conditions were designing process, synthesis tools, and FPGA devices. Further, CLBs for all the devices were assumed to be similar. Additionally, hardware resources are calculated in terms of LUTs, flip-flop, and slices specified in all FPGA devices with different configurations.

5.4 Results and discussions

Power dissipation is an important factor for most of the active RFID tags. To get an optimum balance between the required power and area in active RFID tags, parallel architecture can be preferred. However, this implementation is less energy-efficient and has smaller throughput. However, loop unrolled architecture design with multiple of rounds takes huge area and power consumption, but this design takes less number of clock cycles to complete the encryption operation. As required energy is proportional to the latency of the design, hence this implementation has computed better energy efficiency in terms of energy per bit as shown in Table 2. The pipelined

Table 2 FPGA implementation results of lightweight block ciphers.

Designs	Latency (cycles)	Slice	LUTs	Flip-flops	FMax (MHz)	Throughput (Mbps)	Throughput per slice (Mbps/slice)	Total power (mW)	Energy per bit (nJ/bit)
xc6slx16-3csg324 (Spartan-6)									
PRESENT80 [78]	133	48	170	153	257.40	123.86	2.58	21.61	0.698
PRESENT128 [78]	136	61	220	201	210.66	99.13	1.62	21.76	0.878
Design_1	12	625	1874	768	276.48	1474.56	2.36	20.0	0.014
Design_2	12	82	318	132	255.90	1364.80	16.64	20.0	0.015
Design_3	1	845	2564	234	32.12	2055.68	3.43	20.0	0.009
xc5vlx50t-3ff1136 (Virtex-5)									
PRESENT80 [78]	133	67	190	153	542.30	260.96	3.89	562.75	8.626
PRESENT128 [78]	136	73	239	201	431.78	203.19	2.78	562.67	11.077
Design_1	12	890	2482	768	452.65	2414.13	2.71	568	0.235
Design_2	12	178	428	132	407.95	2175.73	12.22	563	0.259
Design_3	1	938	2779	128	48.78	3121.92	3.32	565	0.181
xc4vlx25-12ff668 (Virtex-4)									
PRESENT80 [78]	133	124	215	153	375.66	180.77	1.46	245.78	5.439
PRESENT128 [78]	136	152	265	201	364.56	171.56	1.13	248.02	5.783
Design_1	12	1508	2921	768	445.13	2374.03	1.57	262	0.110
Design_2	12	322	584	132	342.88	1828.65	5.68	243	0.133
Design_3	1	1719	3260	145	43.61	2791.04	1.62	247	0.088

Continued

Table 2 FPGA implementation results of lightweight block ciphers.—*cont'd*

Designs	Latency (cycles)	Slice	LUTs	Flip-flops	FMax (MHz)	Throughput (Mbps)	Throughput per slice (Mbps/slice)	Total power (mW)	Energy per bit (nJ/bit)
xc3s50-5 (Spartan-3)									
LED80 [79]	32	198	379	74	87.63	175.30	0.88	–	–
LED128 [79]	48	227	444	76	87.63	116.54	0.51	–	–
PRESENT128 [80]	256	117	159	114	114.80	28.46	0.24	–	–
HIGHT [80]	160	91	132	25	163.70	65.48	0.72	–	–
AES128 [81]	534	393	531	338	–	15.86	0.04	23.83	–
AES128 [82]	160	184	–	–	45.64	36.50	0.20	–	–
xTEA [83]	112	254	–	–	62.60	35.78	0.14	–	–
SIMON [84]	–	36	72	–	136	3.60	0.10	–	–

architecture increases the operating frequency and throughput in hardware implementation with acceptable hardware area increment.

A common HF (13.56 MHz) standard is commonly used for credit cards and non-contact smart payment [85]. Therefore, all the proposed architectures are compatible at 13.56 *MHz* frequency for these applications. For a fair comparison, implementations are performed only for encryption version of each cipher.

Three proposed architectures of KLEIN cipher are implemented based on hardware metrics and are realized using synthesis tools. These designs are renamed as follows:

- *Design_1:* A pipelining-based architecture is configured for achieving better performance in terms of operating frequency as shown in Fig. 4. In this design the pipelined registers are placed at the starting of each round. Total 12 registers are utilized to store intermediate states for 64-bit key size.
- *Design_2:* Parallel architecture is designed by sharing the same round function for all the rounds as shown in Fig. 5. The building blocks of round function are reused to process the data. Proposed design completes one round in only one clock cycle and therefore enhances the performance in throughput per slice metric.
- *Design_3:* Loop unrolled design strategy is beneficial for reducing the execution time of a system as shown in Fig. 6. This architecture helps to design an efficient hardware implementation and reduces the energy per bit.

The results of proposed deigns are compared with the state-of-the-art lightweight algorithms and are tabulated in Table 2. The list of state-of-the-art lightweight ciphers in Table 2 includes PRESENT, HIGHT, LED, xTEA, SIMON, and conventional block cipher like AES. The comparisons of the results are also depicted in Figs. 7–9. It can be observed from Table 2 that the proposed designs are better in

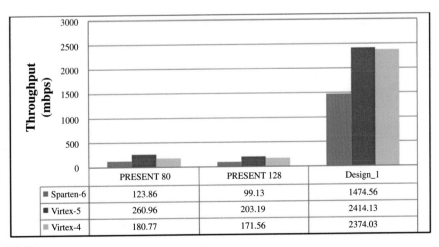

	PRESENT 80	PRESENT 128	Design_1
■ Sparten-6	123.86	99.13	1474.56
■ Virtex-5	260.96	203.19	2414.13
■ Virtex-4	180.77	171.56	2374.03

FIG. 7

Hardware performance of the proposed pipelined architecture for KLEIN block cipher in terms of throughput metric.

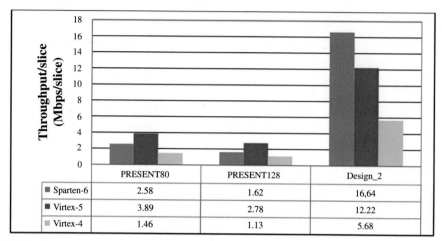

FIG. 8

Hardware performance of proposed parallel architecture for KLEIN block cipher in terms of throughput per slice metric.

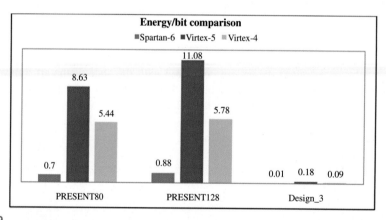

FIG. 9

Hardware performance of proposed unrolled architecture for KLEIN block cipher in terms of energy per bit metric.

terms of latency, maximum frequency, throughput, and energy per bit parameters for different devices in FPGA.

It can be observed from Table 2 and Fig. 7 that the proposed pipelined architecture has improved the operating frequency and throughput metrics. Infect, as per the results, the pipelined architecture results in highest operational frequency among all the proposed and other recent ciphers. The proposed parallel architecture reduces the cost by achieving less hardware area and high throughput (throughput per slice) compared with other lightweight block ciphers as depicted in Fig. 8. Finally the proposed loop unrolled parallel design strategy has resulted in energy efficient design. As shown in Fig. 9, this architecture requires least amount of energy per bit among all the listed ciphers.

6 Performance comparison of conventional and lightweight cryptographic algorithms for IoT

Although the conventional cryptographic algorithms provide high degree of security, it does not make sense to employ these algorithms for resource-constrained applications. These algorithms may use 128 bits or more than 128-bit key, and in turn, it provides high security depending on key length. Utilizing smaller key length can save significant amount of chip area [32]. However, using smaller key length in cryptographic algorithms for resource-constrained applications provides only moderate security. Area requirement for recourse-constrained RFID tags consumes approximately 1000–10,000 gate equivalents (GEs) in ASIC implementation [86], but the conventional cryptographic ciphers, such as DES and AES, require thousands of gates in their hardware implementation. Moreover the software implementations of AES cipher can be fast, but it lays large and complex computations. Therefore conventional cryptographic algorithms are well out of reach for resource-limited applications.

Aforementioned factors play significant roles in the development of modern lightweight algorithms. These algorithms are required to be light for their usability on a vast scope of software and hardware implementations. For example, various low-bit designs like 4-, 8-, 16-, 32-bit processors in software implementation and FPGA and ASIC realizations in hardware implementation.

The performances of conventional and lightweight block ciphers are compared in two different application scenarios. The first scenario includes the transmission of encrypted information for IoT applications, and second is to identify and control the objects by enabling simple response authentication protocol. To accomplish these two application scenarios, lightweight block ciphers are generally used with their two variants, namely, high-speed variant and a low-area or low-memory variant. The former seeks an optimized implementation that can offer small area and memory overhead over speed. On the other hand the later targets the implementation that can improve performance with the cost of additional area and memory footprint. The proposed parallel implementation of KLEIN cipher achieves low area requirement that is possible in design by sharing the same round function for each round. This type of strategy is compatible with limited resource applications such as IoT, RFID, and WSN. However, some applications may not afford to work with these slow response algorithms. Hence, design strategies like loop unrolling and pipelining target the development of high-speed architecture. Here the speed is measured in terms of operating frequency and throughput. High-speed architectures are useful, where a large number of objects have to identify at the same time. The examples of such high-speed applications include tracking, logistics, e-toll, libraries, and shipping. The proposed pipelined and unrolled architectures of KLEIN cipher have better throughput for these applications. Therefore these implementations fulfill the demands of IoT requirements and form the benchmark of lightweight block ciphers.

7 Conclusion

This chapter described the requirements to use lightweight block ciphers for resource-constrained IoT devices and sensor networks. IoT technology has been adapted in several sectors such as logistics, smart city, smart healthcare, automation, transportation, and retail. IoT contains embedded devices that have limitations in terms of size, energy, and memory. The conventional cryptographic algorithms are not suitable for such applications. In this chapter, different lightweight cryptographic algorithms are discussed briefly. Also in this chapter, three architectures of KLEIN lightweight block cipher are proposed. The proposed architectures provide security with high throughput and energy-efficient architectures. In these architectures, pipelined architecture design gives high operating frequency, and high throughput and parallel architecture leads to a good hardware metric in throughput per slice, whereas third one reflects in the reduction of energy per bit parameter. The performance of the proposed architectures can be further enhanced by optimizing the designs.

References

[1] J. Gubbi, R. Buyya, S. Marusic, M. Palaniswami, Internet of things (IoT): a vision, architectural elements, and future directions, Futur. Gener. Comput. Syst. 29 (7) (2013) 1645–1660.
[2] R. Want, S. Dustdar, Activating the internet of things [guest editors' introduction], Computer 48 (9) (2015) 16–20.
[3] H. Suo, J. Wan, C. Zou, J. Liu, Security in the internet of things: a review, in: 2012 International Conference on Computer Science and Electronics Engineering, vol. 3, March 23, IEEE, 2012, pp. 648–651.
[4] G. Ho, D. Leung, P. Mishra, A. Hosseini, D. Song, D. Wagner, Smart locks: lessons for securing commodity internet of things devices, in: Proceedings of the 11th ACM on Asia Conference on Computer and Communications Security, May 30, 2016, pp. 461–472.
[5] D. Airehrour, J. Gutierrez, S.K. Ray, Secure routing for internet of things: a survey, J. Netw. Comput. Appl. 66 (2016) 198–213.
[6] D. Miorandi, S. Sicari, F. De Pellegrini, I. Chlamtac, Internet of things: vision, applications and research challenges, Ad Hoc Netw. 10 (7) (2012) 1497–1516.
[7] L. Da Xu, Enterprise systems: state-of-the-art and future trends, IEEE Trans. Ind. Informatics 7 (4) (2011) 630–640.
[8] Y. Li, M. Hou, H. Liu, Y. Liu, Towards a theoretical framework of strategic decision, supporting capability and information sharing under the context of internet of things, Inf. Technol. Manag. 13 (4) (2012) 205–216.

[9] Z. Pang, Q. Chen, J. Tian, L. Zheng, E. Dubrova, Ecosystem analysis in the design of open platform-based in-home healthcare terminals towards the internet-of-things, in: 15th International Conference on Advanced Communications Technology (ICACT), January 27, IEEE, 2013, pp. 529–534.

[10] S. Misra, M. Maheswaran, S. Hashmi, Security Challenges and Approaches in Internet of Things, Springer International Publishing, Cham, 2017.

[11] M.C. Domingo, An overview of the internet of things for people with disabilities, J. Netw. Comput. Appl. 35 (2) (2012) 584–596.

[12] W. Qiuping, Z. Shunbing, D. Chunquan, Study on key technologies of internet of things perceiving mine, Procedia Eng.. 26 (2011) 2326–2333.

[13] H. Zhou, B. Liu, D. Wang, Design and research of urban intelligent transportation system based on the internet of things, in: Internet of Things, Springer, Berlin, Heidelberg, 2012, pp. 572–580.

[14] B. Karakostas, A DNS architecture for the internet of things: a case study in transport logistics, Procedia Comput. Sci. 19 (2013) 594–601.

[15] B. Koziel, R. Azarderakhsh, M.M. Kermani, A high-performance and scalable hardware architecture for isogeny-based cryptography, IEEE Trans. Comput. 67 (11) (2018) 1594–1609.

[16] P.K. Meher, X. Lou, Low-latency, low-area, and scalable systolic-like modular multipliers for GF (2^m) based on irreducible all-one polynomials, IEEE Trans. Circuits Syst. Regul. Pap. 64 (2) (2016) 399–408.

[17] R. Salarifard, S. Bayat-Sarmadi, H. Mosanaei-Boorani, A low-latency and low-complexity point multiplication in ECC, IEEE Trans. Circuits Syst. Regul. Pap. 65 (9) (2018) 2869–2877.

[18] W. Stallings, Cryptography and Network Security: Principles and Practice, International Edition: Principles and Practice, Pearson Higher Ed, 2014.

[19] FIPS P. 197: Advanced Encryption Standard (AES), National Institute of Standards and Technology, 2001, 26.

[20] K. McKay, L. Bassham, M. Sönmez Turan, N. Mouha, Report on lightweight cryptography, National Institute of Standards and Technology, 2016.

[21] L. Atzori, A. Iera, G. Morabito, The internet of things: a survey, Comput. Netw. 54 (15) (2010) 2787–2805.

[22] A. Whitmore, A. Agarwal, L. Da Xu, The internet of things—a survey of topics and trends, Inf. Syst. Front. 17 (2) (2015) 261–274.

[23] A. Al-Fuqaha, M. Guizani, M. Mohammadi, M. Aledhari, M. Ayyash, Internet of things: a survey on enabling technologies, protocols, and applications, IEEE Commun. Surv. Tutorials 17 (4) (2015) 2347–2376.

[24] A. Botta, W. De Donato, V. Persico, A. Pescapé, Integration of cloud computing and internet of things: a survey, Futur. Gener. Comput. Syst. 56 (2016) 684–700.

[25] R. Davis, The data encryption standard in perspective, IEEE Commun. Soc. Mag. 16 (6) (1978) 5–9.

[26] B. Buhrow, P. Riemer, M. Shea, B. Gilbert, E. Daniel, Block cipher speed and energy efficiency records on the MSP430: system design trade-offs for 16-bit embedded applications, in: International Conference on Cryptology and Information Security in Latin America, September 17, Springer, Cham, 2014, pp. 104–123.

[27] A. Moradi, A. Poschmann, S. Ling, C. Paar, H. Wang, Pushing the limits: a very compact and a threshold implementation of AES, in: Annual International Conference on the Theory and Applications of Cryptographic Techniques, May 15, Springer, Berlin, Heidelberg, 2011, pp. 69–88.

[28] M. Katagi, S. Moriai, Lightweight Cryptography for the Internet of Things, 2008 Sony Corporation, 2008, pp. 7–10.

[29] M. Ebrahim, S. Khan, U.B. Khalid, Symmetric algorithm survey: a comparative analysis, arXiv (2014) preprint arXiv:1405.0398.

[30] D. Dinu, Y. Le Corre, D. Khovratovich, L. Perrin, J. Großschädl, A. Biryukov, Triathlon of lightweight block ciphers for the internet of things, J. Cryptogr. Eng. 9 (3) (2019) 283–302.

[31] R. Benadjila, J. Guo, V. Lomné, T. Peyrin, Implementing lightweight block ciphers on x86 architectures, in: International Conference on Selected Areas in Cryptography, August 14, Springer, Berlin, Heidelberg, 2013, pp. 324–351.

[32] R. Beaulieu, D. Shors, J. Smith, S. Treatman-Clark, B. Weeks, L. Wingers, SIMON and SPECK: block ciphers for the internet of things, in: IACR Cryptology ePrint Archive, 2015, p. 585.

[33] D. Hong, J. Sung, S. Hong, J. Lim, S. Lee, B.S. Koo, C. Lee, D. Chang, J. Lee, K. Jeong, H. Kim, HIGHT: a new block cipher suitable for low-resource device, in: International Workshop on Cryptographic Hardware and Embedded Systems, October 10, Springer, Berlin, Heidelberg, 2006, pp. 46–59.

[34] W. Wu, L. Zhang, LBlock: a lightweight block cipher, in: International Conference on Applied Cryptography and Network Security, June 7, Springer, Berlin, Heidelberg, 2011, pp. 327–344.

[35] M. Izadi, B. Sadeghiyan, S.S. Sadeghian, H.A. Khanooki, MIBS: a new lightweight block cipher, in: International Conference on Cryptology and Network Security, December 12, Springer, Berlin, Heidelberg, 2009, pp. 334–348.

[36] K. Shibutani, T. Isobe, H. Hiwatari, A. Mitsuda, T. Akishita, T. Shirai, Piccolo: an ultra-lightweight blockcipher, in: International Workshop on Cryptographic Hardware and Embedded Systems, September 28, Springer, Berlin, Heidelberg, 2011, pp. 342–357.

[37] L. Li, B. Liu, H. Wang, QTL: a new ultra-lightweight block cipher, Microprocess. Microsyst. 45 (2016) 45–55.

[38] W. Zhang, Z. Bao, D. Lin, V. Rijmen, B. Yang, I. Verbauwhede, RECTANGLE: a bitslice lightweight block cipher suitable for multiple platforms, SCIENCE CHINA Inf. Sci. 58 (12) (2015) 1–5.

[39] L. Knudsen, G. Leander, A. Poschmann, M.J. Robshaw, PRINTcipher: a block cipher for IC-printing, in: International Workshop on Cryptographic Hardware and Embedded Systems, August 17, Springer, Berlin, Heidelberg, 2010, pp. 16–32.

[40] J. Borghoff, A. Canteaut, T. Güneysu, E.B. Kavun, M. Knezevic, L.R. Knudsen, G. Leander, V. Nikov, C. Paar, C. Rechberger, P. Rombouts, PRINCE–a low-latency block cipher for pervasive computing applications, in: International Conference on the Theory and Application of Cryptology and Information Security, December 2, Springer, Berlin, Heidelberg, 2012, pp. 208–225.

[41] C.H. Lim, T. Korkishko, mCrypton–a lightweight block cipher for security of low-cost RFID tags and sensors, in: International Workshop on Information Security Applications, August 22, Springer, Berlin, Heidelberg, 2005, pp. 243–258.

[42] T. Suzaki, K. Minematsu, S. Morioka, E. Kobayashi, Twine: a lightweight, versatile block cipher, in: ECRYPT Workshop on Lightweight Cryptography, November 28, 2011 2011.

[43] A. Bogdanov, L.R. Knudsen, G. Leander, C. Paar, A. Poschmann, M.J. Robshaw, Y. Seurin, C. Vikkelsoe, PRESENT: an ultra-lightweight block cipher,

in: International Workshop on Cryptographic Hardware and Embedded Systems, September 10, Springer, Berlin, Heidelberg, 2007, pp. 450–466.

[44] L. Li, B. Liu, Y. Zhou, Y. Zou, SFN: a new lightweight block cipher, Microprocess. Microsyst. 60 (2018) 138–150.

[45] T.P. Berger, J. Francq, M. Minier, G. Thomas, Extended generalized Feistel networks using matrix representation to propose a new lightweight block cipher: Lilliput, IEEE Trans. Comput. 65 (7) (2015) 2074–2089.

[46] Z. Gong, S. Nikova, Y.W. Law, KLEIN: a new family of lightweight block ciphers, in: International Workshop on Radio Frequency Identification: Security and Privacy Issues, June 26, Springer, Berlin, Heidelberg, 2011, pp. 1–18.

[47] S. Banik, A. Bogdanov, T. Isobe, K. Shibutani, H. Hiwatari, T. Akishita, F. Regazzoni, Midori: a block cipher for low energy, in: International Conference on the Theory and Application of Cryptology and Information Security, November 29, Springer, Berlin, Heidelberg, 2015, pp. 411–436.

[48] D. Hong, J.K. Lee, D.C. Kim, D. Kwon, K.H. Ryu, D.G. Lee, LEA: a 128-bit block cipher for fast encryption on common processors, in: International Workshop on Information Security Applications, August 19, Springer, Cham, 2013, pp. 3–27.

[49] P. Singh, B. Acharya, R.K. Chaurasiya, A comparative survey on lightweight block ciphers for resource constrained applications, Int. J. High Perform. Syst. Archit. 8 (4) (2019) 250–270.

[50] S. Singh, P.K. Sharma, S.Y. Moon, J.H. Park, Advanced lightweight encryption algorithms for IoT devices: survey, challenges and solutions, J. Ambient. Intell. Humaniz. Comput. 24 (2017) 1–8.

[51] R. Mehta, Distributed denial of service attacks on cloud environment, Int. J. Adv. Res. Comput. Sci. 8 (5) (2017) 2204–2206.

[52] N. Hanley, M. ONeill, Hardware comparison of the ISO/IEC 29192-2 block ciphers, in: 2012 IEEE Computer Society Annual Symposium on VLSI, August 19, IEEE, 2012, pp. 57–62.

[53] D. Coppersmith, R. Gennaro, S. Halevi, C.S. Jutla, S.M. Matyas Jr., L.J. O'connor, M. Peyravian, D.R. Safford, N. Zunic, inventors; International Business Machines Corp, assignee, Symmetric block cipher using multiple stages with modified type-1 and type-3 feistel networks. United States Patent 6,189,095, (2001).

[54] G. Leander, C. Paar, A. Poschmann, K. Schramm, New lightweight DES variants, in: International Workshop on Fast Software Encryption, March 26, Springer, Berlin, Heidelberg, 2007, pp. 196–210.

[55] R. Beaulieu, D. Shors, J. Smith, S. Treatman-Clark, B. Weeks, L. Wingers, The SIMON and SPECK lightweight block ciphers, in: Proceedings of the 52nd Annual Design Automation Conference, June 7, 2015, pp. 1–6.

[56] C. De Canniere, O. Dunkelman, M. Knežević, KATAN and KTANTAN—a family of small and efficient hardware-oriented block ciphers, in: International Workshop on Cryptographic Hardware and Embedded Systems, September 6, Springer, Berlin, Heidelberg, 2009, pp. 272–288.

[57] M. Robshaw, O. Billet, New Stream Cipher Designs: The eSTREAM Finalists, Springer, 2008.

[58] B.K. Roy, W. Meier, Fast Software Encryption, 11th International Workshop, FSE, 2004, pp. 5–7.

[59] Iwata T, Cheon JH, editors. Advances in Cryptology—ASIACRYPT 2015: 21st International Conference on the Theory and Application of Cryptology and Information Security, Auckland, New Zealand, November 29–December 3, 2015, Proceedings. Springer; 2015.

[60] V.A. Ghafari, H. Hu, C. Xie, Fruit: ultralightweight stream cipher with shorter internal state, eSTREAM: ECRYPT Stream Cipher Project, (2016).

[61] M. Hamann, M. Krause, W. Meier, Lizard—a lightweight stream cipher for power-constrained devices, IACR Trans. Symm. Cryptol. 8 (2017) 45–79.

[62] M. Feldhofer, C. Rechberger, A case against currently used hash functions in RFID protocols, in: OTM Confederated International Conferences" on the Move to Meaningful Internet Systems, October 29, Springer, Berlin, Heidelberg, 2006, pp. 372–381.

[63] A. Bogdanov, G. Leander, C. Paar, A. Poschmann, M.J. Robshaw, Y. Seurin, Hash functions and RFID tags: mind the gap, in: International Workshop on Cryptographic Hardware and Embedded Systems, August 10, Springer, Berlin, Heidelberg, 2008, pp. 283–299.

[64] S. Hirose, K. Ideguchi, H. Kuwakado, T. Owada, B. Preneel, H. Yoshida, A lightweight 256-bit hash function for hardware and low-end devices: lesamnta-LW, in: International Conference on Information Security and Cryptology, December 1, Springer, Berlin, Heidelberg, 2010, pp. 151–168.

[65] A. Bogdanov, M. Knežević, G. Leander, D. Toz, K. Varıcı, I. Verbauwhede, SPONGENT: a lightweight hash function, in: International Workshop on Cryptographic Hardware and Embedded Systems, September 28, Springer, Berlin, Heidelberg, 2011, pp. 312–325.

[66] J. Guo, T. Peyrin, A. Poschmann, The PHOTON family of lightweight hash functions, in: Annual Cryptology Conference, August 14, Springer, Berlin, Heidelberg, 2011, pp. 222–239.

[67] J.P. Aumasson, L. Henzen, W. Meier, M. Naya-Plasencia, Quark: a lightweight hash, J. Cryptol. 26 (2) (2013) 313–339.

[68] P. Rogaway, T. Shrimpton, Cryptographic hash-function basics: definitions, implications, and separations for preimage resistance, second-preimage resistance, and collision resistance, in: International Workshop on Fast Software Encryption, February 5, Springer, Berlin, Heidelberg, 2004, pp. 371–388.

[69] T. Grembowski, R. Lien, K. Gaj, N. Nguyen, P. Bellows, J. Flidr, T. Lehman, B. Schott, Comparative analysis of the hardware implementations of hash functions SHA-1 and SHA-512, in: International Conference on Information Security, September 30, Springer, Berlin, Heidelberg, 2002, pp. 75–89.

[70] J.S. Cho, S.S. Yeo, S.K. Kim, Securing against brute-force attack: a hash-based RFID mutual authentication protocol using a secret value, Comput. Commun. 34 (3) (2011) 391–397.

[71] J. Daemen, V. Rijmen, A new MAC construction ALRED and a specific instance ALPHA-MAC, in: International Workshop on Fast Software Encryption, February 21, Springer, Berlin, Heidelberg, 2005, pp. 1–17.

[72] M. Dworkin, Recommendation for block cipher modes of operation. Methods and techniques, National Inst of Standards and Technology, Gaithersburg MD, 2001.

[73] Z. Gong, P. Hartel, S. Nikova, S.H. Tang, B. Zhu, TuLP: a family of lightweight message authentication codes for body sensor networks, J. Comput. Sci. Technol. 29 (1) (2014) 53–68.

[74] M. Girault, R. Cohen, M. Campana, A generalized birthday attack, in: Workshop on the Theory and Application of Cryptographic Techniques, May 25, Springer, Berlin, Heidelberg, 1988, pp. 129–156.

[75] B.J. Mohd, T. Hayajneh, A.V. Vasilakos, A survey on lightweight block ciphers for low-resource devices: comparative study and open issues, J. Netw. Comput. Appl. 58 (2015) 73–93.

[76] J. Aragones-Vilella, A. Martinez-Balleste, A. Solanas, A brief survey on RFID privacy and security, in: World Congress on Engineering, July, 2007, pp. 1488–1493.

[77] A. Jules, A research survey: RFID security and privacy issue, Comput. Sci. 24 (2006) 381–394.

[78] C.A. Lara-Nino, A. Diaz-Perez, M. Morales-Sandoval, Lightweight hardware architectures for the present cipher in FPGA, IEEE Trans. Circuits Syst. Regul. Pap. 64 (9) (2017) 2544–2555.

[79] N.N. Anandakumar, T. Peyrin, A. Poschmann, A very compact FPGA implementation of LED and PHOTON, in: International Conference on Cryptology in India, December 14, Springer, Cham, 2014, pp. 304–321.

[80] P. Yalla, J.P. Kaps, Lightweight cryptography for FPGAs, in: 2009 International Conference on Reconfigurable Computing and FPGAs, December 9, IEEE, 2009, pp. 225–230.

[81] J.P. Kaps, B. Sunar, Energy comparison of AES and SHA-1 for ubiquitous computing, in: International Conference on Embedded and Ubiquitous Computing, August 1, Springer, Berlin, Heidelberg, 2006, pp. 372–381.

[82] J. Chu, M. Benaissa, Low area memory-free FPGA implementation of the AES algorithm, in: 22nd International Conference on Field Programmable Logic and Applications (FPL), August 29, IEEE, 2012, pp. 623–626.

[83] J.P. Kaps, Chai-tea, cryptographic hardware implementations of xtea, in: International Conference on Cryptology in India, December 14, Springer, Berlin, Heidelberg, 2008, pp. 363–375.

[84] A. Aysu, E. Gulcan, P. Schaumont, SIMON says: break area records of block ciphers on FPGAs, IEEE Embed. Syst. Lett. 6 (2) (2014) 37–40.

[85] S. Rizkalla, R. Prestros, C.F. Mecklenbräuker, Optimal card design for non-linear HF RFID integrated circuits with guaranteed standard-compliance, IEEE Access 6 (2018) 47843–47856.

[86] A. Juels, S.A. Weis, Authenticating pervasive devices with human protocols, in: Annual International Cryptology Conference, August 14, Springer, Berlin, Heidelberg, 2005, pp. 293–308.

EELC: Energy-efficient lightweight cryptography for IoT networks

G. Rajesh[a], X. Mercilin Raajini[b], K. Martin Sagayam[c], A. Sivasangari[d], and Lawrence Henesey[e]

[a]*Department of Information Technology, MIT Campus, Anna University, Chennai, India*
[b]*Department of ECE, Prince Shri Venkateshwara Padmavathy Engineering College, Chennai, India* [c]*Department of ECE, Karunya Institute of Technology and Sciences, Coimbatore, India* [d]*School of Computing, Sathyabama Institute of Science and Technology, Chennai, India* [e]*Department of CSE, Blekinge Institute of Technology, Karlskrona, Sweden*

Chapter outline

Security and privacy issues in IoT devices and sensor networks. https://doi.org/10.1016/B978-0-12-821255-4.00009-2

1 Introduction—IoTSec

Internet of Things security (IoTSec) is a challenging research area in IoT-related network technologies due to various constraints. It has been predicted that 32 million IoT devices will be deployed for various applications in the world by 2020. IoT devices are being penetrated in all kind of physical entities (such as smart home, smart industry, smart city, hospital, power grid, traffic control, and wearable devices). Nowadays, IoT devices are more personal to human lives, which sense, process, and communicate the sensitive, critical, private, and more intimate personal data from one end to another. It is now believed that considerations of IoT security robustness have become a significant part of IoT research due to the continuously increasing number of threats, vulnerabilities, and bad characters attacking IoT data and network.

Cybersecurity or Information technology security (IT security) is the protection of computer systems from the damage or theft of hardware, software, or electronic data and from the disruption or misdirection of the services they provide. Information security is to ensure aspects such as data confidentiality, data integrity, authentication, and nonrepudiation by using modern cryptography. Cybersecurity is the technique of protecting computers, networks, programs, and data from unauthorized access. Network security includes activities to protect the usability, reliability, integrity, and safety of the network. The major divisions of various levels of cybersecurity are critical infrastructure security, application security, network security, cloud security, and Internet of things (IoT) security.

1.1 Cyberattacks

A cyberattack is an intentional exploitation of computer systems and related devices with malicious code or process to alter the data or information in ICT applications and networks. This offensive action will target computer information systems, network peripherals, communication devices, and network infrastructure to alter or destroy or steal data or information locally or remotely.

Denial of service (DoS) and distributed denial-of-service (DDoS) attacks

- Man-in-the-middle (MITM) attack
- Phishing and spear-phishing attacks
- Drive-by attack
- Birthday attack
- Malware attack
- Brute force attack and password attack
- SQL injection attack
- Cross-site scripting (XSS) attack
- Eavesdropping attack

Possible security attacks in a data communication system are illustrated in Fig. 1. IoT security is freedom from or resilience against the potential harm caused by others to the IoT network's devices, data generated and processed by it.

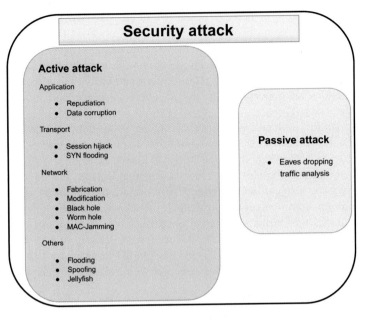

FIG. 1

Security attacks in communication systems.

1.2 Security challenges in IoT

The IoT architecture is represented by three-layer architecture, namely, (1) sensing layer or perception layer, (2) network or transport layer, and (3) application layer. The sensing layer consists of low-power IoT devices that are responsible for acquiring data from the physical environment and share it to the user through the communication network. If the attacker gets the privilege to control these tiny devices, they could easily access all the data generated by these devices. Apart from data theft the devices can be easily tampered physically or even replaced with some other malicious devices. The primary security attacks that come from the perception layer are physical attacks by node tampering or malicious code injection, impersonation, data transit attacks, routing attacks like sniffing and man-in-the-middle attack, and denial-of-service attacks (DoS).

The second layer or transport layer depends on the existing core Internet infrastructure like Wi-Fi, Bluetooth, 3G, and ad hoc networks to transfer the data from the perception layer to the application layer. This communication layer is more vulnerable to attackers or intruders like DoS attack, routing attack, and data transit attacks. The application layer includes a storage area like cloud or server or any storage unit to keep the data sensed by the IoT devices for analytics and processing. The application layer responds to the user's query like giving the temperature or weather prediction based on the data sensed by the IoT devices. This layer is vulnerable to data leakage, data transit attack, and DoS attack.

The diverse nature of heterogeneous IoT devices is now being connected to the network in various applications like Internet of Industrial Things (IoIT), Internet of Medical Things (IoMT), Internet of Drone Things (IoDT), and Internet of Social Things (IoST). The devices and protocols followed by them all are developed by different vendors, which require interoperability, and also, this diverse nature provides an opportunity to the intruders to attack and manipulate the IoT system.

Authorization, authentication, privacy, and data confidentiality are major security issues in IoT networks. Outside attackers attack the network by stealing the security key or parameters mirrored in the IoT devices. By using those keys the attacker can intrude the network, recreate a virtual device, and inject false data to the server and also can access the secure information from the server. This type of attack in the communication networks is called a side-channel attack. In side-channel attacks the intruder will steal the keys of the security algorithms used. Using these keys, attackers will hack the server. Various works of literature explored the increased amount of DDoS attack in IoT devices. An IoT device with a proper cryptosystem that includes a secured key exchange security mechanism can avoid such situations.

Researchers have been working on these issues and proposed many solutions to prevent a cyberattack. Crosby and Varadarajan validated the IoT security by enhancing the end-to-end security in 6LoWPAN. Granjal and his team tested various crypto techniques like SHA1, SHA2, DES, and AES on low-power sensor nodes to validate the trustworthiness of the network. IoT networks are self-configurable and autoconfigurable; it means that IoT networks have the flexibility of nodes joining any time to the network. Those nodes should be automatically authenticated and authorized by the network to ensure the trustworthiness.

Based on the significance of IoT security and its challenges, we have identified the following open issues for future research:

(i) There is a significant need for authentication and authorization for IoT end device.
(ii) Privacy for data generated by the critical IoT end system.
(iii) End-to-end data security with better cryptosystem.
(iv) Enforcement of security standardization in the network.

To overcome the earlier research gap, here, we proposed a novel security algorithm Energy-Efficient Lightweight Cryptography (EELC) for low-power devices shown in Fig. 2.

Three lightweight security techniques:

(i) energy-efficient subkey generation (EESG) method,
(ii) lightweight MAC generation function (LMGF),
(iii) energy-efficient encryption (EEE) algorithm.

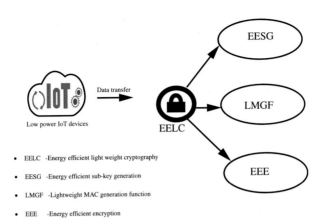

Low power IoT devices

Data transfer

EELC

EESG

LMGF

EEE

- EELC -Energy efficient light weight cryptography
- EESG -Energy efficient sub-key generation
- LMGF -Lightweight MAC generation function
- EEE   -Energy efficient encryption

FIG. 2

Scope diagram of proposed EELC.

to maintain security without compromising energy and processing power consumptions have been formulated. The performance of EELC is evaluated with existing RC5 and SHA algorithms.

The rest of the sections is as follows: Section 2 provides the literature survey highlighting the work related to EELC. Section 3 gives the details about the proposed work, contribution, overall architecture, and mathematical model. Section 4 discusses the results of performance metrics. Section 5 contains the conclusion and future work that can be carried out.

2 Literature survey

Cryptographic algorithms for providing security to the data based on various parameters are still facing various challenges toward meeting security with minimal power requirements and time consumption. Various symmetric and asymmetric algorithms are proposed to meet the challenges. In our proposed work, security constraints have been analyzed with basic algorithms like SHA and RC1. Secure hash algorithm (SHA) is a hash function/cryptographic function that provides security to the input data, which can be mathematically encrypted to improve the security. Message authentication, digital signature, and key generation will enrich the security of the incoming data. RC5, a symmetric key block cipher stands for Rivest Cipher, encrypts the data and provides security based on key factors like block size, key size, and the number of rounds. Power consumption and time consumption are the main challenges faced during encryption and decryption. Based on the limitations observed in existing algorithms, the proposed Energy-Efficient Lightweight Cryptography algorithm is modeled such that the speed of execution and power consumption is optimized.

3 Energy-efficient lightweight cryptography (EELC) architecture

Proposed Energy-Efficient Lightweight Cryptography (EELC) consisting of three specialized algorithms

(i) subkey generation
(ii) MAC generation
(iii) encryption

Fig. 3 shows the architecture of the proposed algorithms (EESG, LMGF, and EEE). It shows the sharing of the same component in MAC generation, encryption, and decryption process. The plain text in the figure refers to the input data. (i) The subkeys are generated using the EESG subkey generation process followed by the message authentication code (MAC) generation using LMGF using the generated subkeys. (iii) The plain text, along with the MAC, is encrypted using EEE.

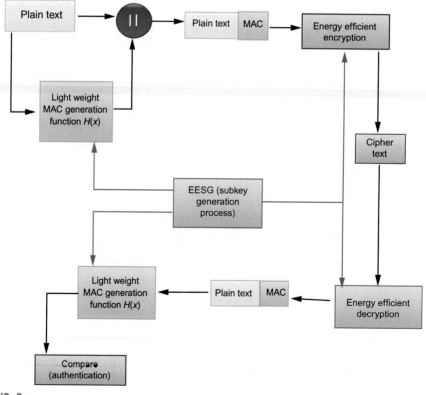

FIG. 3

Architecture of energy-efficient lightweight cryptography.

(iv) The encrypted message is sent to the receiver's end, where it is decrypted using the energy-efficient decryption algorithm. The resultant is the plain text along with the MAC. (v) The receiver generates subkeys using the EES, followed by the MAC generation using the LMG. (vi) Finally the received MAC and the MAC generated at the receiver's end are compared with permit authentication. This process ensures that there is no loss in the original data. In this way, secure communication is established.

3.1 Energy efficient subkey generation

EESG works on bits or binary numbers—the 0 s and 1 s common to digital computers. EESG uses a 128-bit master key to generate eight subkeys each of 64 bit. A permuted key is used, which is of 64-bit length. This permuted key is initialized randomly. These subkeys are used in MAC generation and also encryption. It is shown in Fig. 4.

The master key is split into eight blocks each consisting of 16 bits of the 128 bits. The permuted key is split into two, 32 bit blocks.

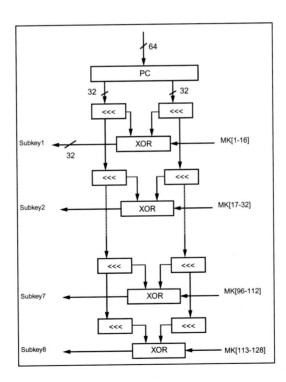

FIG. 4

Working of EESG.

Example. Master key length = 128 bits
0110001101101000011100100110100101011100110111011001100001011011010011
1001101101001011100100110111101110011011010100001100001011011110
    Initial permuted key = 64 bits
0101101011111001110001100110100101011011100001010101011011110010001
    The master key is split into eight blocks each consisting of 16 bits of the 128 bits. For example, consider the earlier master key. It is split into eight blocks as follows:

    MKsplit [0] = 0110001101101000
    MKsplit [1] = 0111001001101001
    MKsplit [2] = 0111001101110110
    MKsplit [3] = 0110000101101101
    MKsplit [4] = 0111001101101001
    MKsplit [5] = 0111001001101111
    MKsplit [6] = 0111001101101000
    MKsplit [7] = 0110000101101110

The permuted key is split into two, 32 bit blocks. For example, consider the earlier permuted key. It splits as follows:

    PKsplit [0] = 01011010111110011100011001101001
    PKsplit [1] = 01110111000010101011011110010001

For each subkey the permuted key is rotated thrice. These blocks are used along with the permuted key for enhancing the strength of the generated subkeys. The working of the subkey generation is shown in Fig. 3. Here, PC represents the permuted constant; MK refers to the master key. Here, in Fig. 3, XOR represents a simple function in which the rotated right of the permuted key is XORed with the master key. The master key split is 16 bits. So, it padded along with zeros.

Example. MKsplit [0] = 0110001101101000

Right padding: 0000000000000000 0110001101101000
Left padding: 0110001101101000 0000000000000000
    The permuted key is rotated three times for the left side, or it is rotated three times for the right side. The earlier steps are repeated for each round. So, there is a better spread of bits.

Example. PKrightsplit [0] = 01011010111110011100011001101001
    PKleftsplit [1]= 01110111000010101011011110010001

After rotating it three times,

    SL0 — 11010111110011100011001101001010
    SR0 = 10111000010101011011110010001011

The left split of the permuted key is XORed with the left padding of master key, and the right split of the permuted key is XORed with the right padding of the master key, and finally the results from the left and right side are concatenated.

Example. Subkey [0] = 11010111110011100101000000100010110110110011110
1101111001001011

For the next subkey the already rotated permuted key is again rotated three times for both the splits, and the same function is carried out along with the second split of the master key. This process goes on until eight subkeys are generated.

Example. SL1 = 10111110011100011001101001010110
SR1 = 11000010101011011110010001011101

Subkey [1] = 101111100111000111010000011111110110000110001001110010
001011101

SL2 = 11110011100011001101001010110101
SR2 = 00010101011011110010001011101110

Subkey [2] = 1111001110001100101000011100001101100110000110010010001
011101110

SL3 = 10011100011001101001010110101111
SR3 = 10101011011110010001011101110000

Subkey [3] = 1001110000110011011111010011000010110010100001010000010111
01110000

SL4 = 11100011001101001010110101111100
SR4 = 01011011110010001011101110000101

Subkey [4] = 11100011001101001101111000010101001010001010000110111011
10000101

SL5 = 00011001101001010110101111100111
SR5 = 11011110010001011101110000101010

Subkey [5]= 00011001101001010001100110001000101011000010101011011100
00101010

SL6 = 11001101001010110101111100111000
SR6 = 11110010001011101110000101010110

Subkey [6] = 11001101001010110010110001010000100000010100011011100001
01010110

SL7 = 01101001010110101111100111000110
SR7 = 10010001011101110000101010110111

Subkey [7] = 01101001010110101001100010101000111100000001100100001010
10110111

The subkeys generated in EELC are used in LMGF and EEE algorithm.

3.2 LMGF algorithm

The MAC is generated using the master key, which is 128 bits as explained earlier, and it also uses the eight subkeys. The generated MAC is 128 bits in length. The algorithm takes the input as a message string with a maximum length of $\leq 2^{128}$ bits. The input is processed in 1024-bit blocks. Two constants a and b are defined, where a is 55 55 55 55 AA AA AA AA and b is AA AA AA AA 55 55 55 55. These constants are named as uniformity constants. The LMGF is carried out, as shown in Fig. 5.

3.2.1 Algorithm

Step 1 Append padding bits. The message is padded so that its length is congruent to 896 modulo 1024 [*Length* ≡ 896 % 1024]. Padding is always added, even if the message is already of the desired length. Thus the number of padding bits is in the range of 1–1024. The padding consists of a single 1 bit followed by the necessary number of 0 bits.

  Step 2 Initialize hash buffer. A 128-bit buffer is used to hold intermediate and final results of the hash function. The buffer can be represented as two 64-bit registers (a,b). These registers are initialized to the following 64-bit integers (hexadecimal values):

$$a = 55\,55\,55\,55\,AA\,AA\,AA\,AA$$

$$b = AA\,AA\,AA\,AA\,55\,55\,55\,55$$

Step 3 Process the message in 1024-bit (128-word) blocks. The heart of the algorithm is a module that consists of eight rounds; this module is shown in Fig. 6. In Fig. 6, W represents each 128-bit split of the 1024-bit input; K represents the subkey. Each

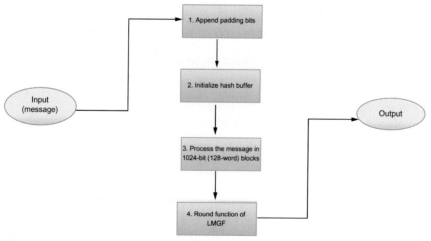

FIG. 5

Flow of LMGF.

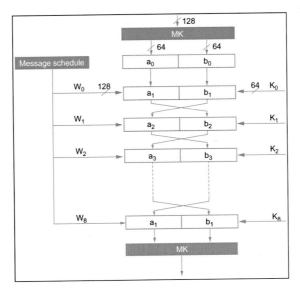

FIG. 6

LMGF processing for a 1024 bit.

subkey is used for its respective round. So a total of eight rounds and a,b represents the uniformity constant as mentioned earlier. Initially the master key is split into two blocks of 64 bits each. These two blocks are added with uniformity constants. The resultant along with 128-bit message digest and the subkey is processed separately and stored in the buffer as an intermediate for the next round.

Step 4 Round function of LMGF. The round function is carried out, as shown in Fig. 7.

Here the uniformity constants are rotated three times; the constant *b* is added along with the subkey and the first 64 bits of the message, which is considered as the resultant one. This resultant one is added along with next 64 bits of the message, which is considered as resultant two. Now the resultant one is assigned to the buffer *a*, and resultant two is assigned to the buffer *b*. This is a process that spreads the data equally.

Step 5 Output. The final 64-bit data in the buffers are added with two blocks the master key again, and finally the data are concatenated, which results in the 128-bit message authentication code.

Fig. 6 represents the round function used in LMGF, and the mathematical model for the round function is explained as follows:

3.3 Mathematical model for MAC

$$T_1 = (b \lll 3) + w_{1-64} + k \tag{1}$$

$$T_2 = T_1 + (a \lll 3) + w_{65-128} \tag{2}$$

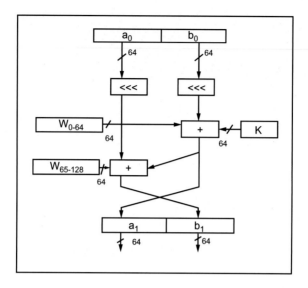

FIG. 7

LMGF round function.

Here, T_1 represents the resultant one, and T_2 represents the resultant two. K represents the subkey, W is the message, and a,b represent the constants. Fig. 8 shows the model for MAC generation, as explained earlier. The mathematical model for the overall MAC is explained as follows:

$$MAC = \sum_{i=0}^{n} (a_{i-1} + t_i + w_i) \bullet t_i \tag{3}$$

$$t_i = b_i + k_i + w_i \tag{4}$$

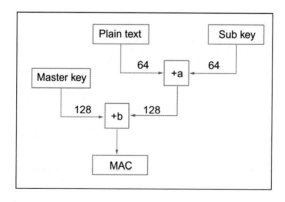

FIG. 8

LMGF model.

3.4 **Flow of OMGF**

This section shows an example of MAC generation using LMGF.

Example. Consider the master key as used previously in subkey generation.
MKlength = 1280110001101101000011100100110100101110011011101101100110 10000101101101011100110110100101110010011011110111001101101010000110 000101101110

This is split into two blocks.
MK1 Length =6401100011011010000111001001101001011100110111011001 10000101101101

MK2 Length = 64011100110110100101110010011011110111001101101010000 110000101101110

Consider a message of 1024 bits.
binaryMessage = 1024
110100001100101011011000110110001101111000.....padded with zeros.

This input message is split into 8128 bits, where each 128-bit block is further split into 2 64 bit blocks.

W [0] =
1101000011001010110110001101100011011111000000000000000000000000000000

W [1] =
0000000000000000000000000000000000000000000000000000000000000000000000

W [2] =
0000000000000000000000000000000000000000000000000000000000000000000000

W [3] =
0000000000000000000000000000000000000000000000000000000000000000000000

W [4] =
0000000000000000000000000000000000000000000000000000000000000000000000

W [5] =
0000000000000000000000000000000000000000000000000000000000000000000000

W [6] =
0000000000000000000000000000000000000000000000000000000000000000000000

W [7] =
0000000000000000000000000000000000000000000000000000000000000000000000

W [8] =
0000000000000000000000000000000000000000000000000000000000000000000000

W [9] =
0000000000000000000000000000000000000000000000000000000000000000000000

W [10] =
0000000000000000000000000000000000000000000000000000000000000000000000

W [11] =
0000000000000000000000000000000000000000000000000000000000000000000000

W [12] =
0000000000000000000000000000000000000000000000000000000000000000000000

W [13] =
00000000000000000000000000000000000000000000000000000000000000000

W [14] =
00000000000000000000000000000000000000000000000000000000000000000

W [15] =
00000000000000000000000000000000000000000000000000000000000000000

These inputs are processed using the round function as explained earlier, and finally, we obtain the MAC.

MAC = 12811000111000000010001001111100111011100110010010101100
00111010000001100000011110000100101110101010100110110111110010001000
1000101

MAC = Ç çs$°è êIÞDE, this is the equivalent MAC.

3.5 Hardware requirement

In Fig. 9 the hardware model for LMFG is shown. It is clear that no complex operations are being used. So the computation complexity is less compared with the universal MAC generation. Simple operators like XOR and left shifter are only used, and so, no high complex operations are required. These operators will make the LMGF as a suitable slight weight MAC generation algorithm.

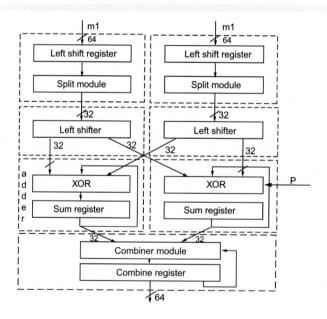

FIG. 9

Hardware requirement for LMGF.

3.6 Robustness of OMGF

On reducing the higher functionalities, the power and strength of the MAC generated might go down. So, here, it is proven that on reducing the higher functionalities, the overall complexity of the MAC generated is not disturbed, and it is as strong as other MAC generating algorithms.

Theorem:

For any even $n \geq 2$ and $8 \leq w \geq 1$, $PH[n,w]$ is 2^{-w}, universally defined on n equal-length strings.

To prove:

$$\Pr[PH_k(M) = PH_k(M')] \leq 2^{-w} \tag{5}$$

Pr	Probability
PH	Polynomial
K	Kth position
M and M'	Distinct messages of equal size
W	Weight

$$\Pr\left[\sum_{i=0}^{n} (a_{i\ 1} + t_i + w_i) \bullet (b_i + k_i + w_i) = \sum_{i=0}^{n} (a_{i\ 1} + t_i + w'_i) \bullet (b_i + k_i + w'_i)\right] \underset{=}{\leq} ?^{-w} \tag{6}$$

The probability is spread uniformly, and the arithmetic is taken over GF (2). We know that M and M' are distinct. So, $w_i \neq w_i'$

$$\Pr[(a_0 + t_0 + w_0) \bullet (b_0 + k_0 + w_0) + (a_1 + t_1 + w_1) \bullet (b_1 + k_1 + w_1)$$
$$+ \sum_{i=2}^{n} (a_{i-1} + t_i + w_i) \bullet (b_i + k_i + w_i) = (a_0 + t_0 + w'_0) \bullet (b_0 + k_0 + w'_0)$$
$$+ (a_1 + t_1 + w'_1) \bullet (b_1 + k_1 + w'_1) + \sum_{i=2}^{n} (a_{i-1} + t_i + w'_i) \bullet (b_i + k_i + w'_i)] \leq 2^{-w} \tag{7}$$

$$Let\ Y = \sum_{i=2}^{n} (a_{i-1} + t_i + w'_i) \bullet (b_i + k_i + w'_i) - \sum_{i=2}^{n} (a_{i-1} + t_i + w_i) \bullet (b_i + k_i + w_i) \tag{8}$$

Using Eq. (8) in Eq. (7), we get

$$\Pr[(a_0 + t_0 + w_0) \bullet (b_0 + k_0 + w_0) + (a_1 + t_1 + w_1) \bullet (b_1 + k_1 + w_1)$$
$$- (a_0 + t_0 + w'_0) \bullet (b_0 + k_0 + w'_0) - (a_1 + t_1 + w'_1) \bullet (b_1 + k_1 + w'_1) = Y] \tag{9}$$

Next, for any w_1 and w_1', there exists at most one key such that

$$((a_0 + t_0) \bullet (b_0 + k_0) (w_1 - w'_1)) + (w_0)(w_1 + (a_1 + t_1) \bullet (b_1 + k_1))$$
$$- (w'_0)(w'_1 + (a_1 + t_1) \bullet (b_1 + k_1)) = Y \tag{10}$$

Then the identity becomes

$$((a_0 + t_0) \bullet (b_0 + k_0)(w_1 - w_1')) = Y - (w_0)(w_1 + (a_1 + t_1) \bullet (b_1 + k_1))$$
$$+ (w_0')(w_1' + (a_1 + t_1) \bullet (b_1 + k_1)) \qquad (11)$$

Since $w_1 \neq w_1'$, LHS can never be zero:

$$(a_0 + t_0) \bullet (b_0 + k_0) = [Y - (w_0)(w_1 + (a_1 + t_1) \bullet (b_1 + k_1))$$
$$+ (w_0')(w_1' + (a_1 + t_1) \bullet (b_1 + k_1))]/(w_1 - w_1') \qquad (12)$$

So, there are two possible cases: we can get either exact divisor or have a divisor with some reminder. Therefore, out of two cases, there can be only one case that causes a collision and breaks the key (out of 2^w possible values).

Therefore

$$\Pr[PH_k(M) = PH_k(M')] \leq 2^{-w}$$

Hence, proved.

So, OMGF is very strong, and the key is not easy to break.

3.7 Energy efficient encryption

Here the encryption technique uses the subkeys generated using EESG and two other techniques flipping of bits and single-point crossover.

Flipping:

If the bit is "0" change it to "1".

Else if the bit is "1" change it to "0."

E.g., 101010100110 ➜ 010101011001.

Single-point crossover:

Cut the data into two, and exchange the sections.

E.g., 100|0000|000 ➜ 000|0000|100.

This encryption technique introduces a shared key that is known to both the sender and receiver. The shared key is a 10-digit number. The sum of all the elements in the shared key gives 1024. The shared key is changed periodically to make the encryption even secure. The input data are split based on the shared key, and other functionalities are added after the split is done. Here the input length is same as OMGF, which is 1024 bits. This will be split based on the shared key (Table 1).

E.g., 6,1,3,6,3,2,6,1,2,0 gives $64 + 128 + 32 + 64 + 32 + 256 + 64 + 128 + 256 + 0 = 1024$.

Consider $n = 0$ initially.

Table 1 Shared key in EEE.

Element in shared key	Value
0	0 bits
3	32 bits
6	64 bits
1	128 bits
2	256 bits

Step 1: Split the input based on the elements in the shared key, called a split string.
Step 2: If the length of the split string is 64, then.
  (i) add a split string with the *n*th subkey
  (ii) increment *n*.
Step 3: If the length of the split string is not 64, then.
  (i) cut the split string into two sections
      (I) first 1/3 of the length of the split string,
      (II) Last 1/3 of the length of the split string,
  (ii) perform flipping of bits for the two sections,
  (iii) cut the first and last 8 bits of the split string and perform crossover.

3.8 Working of EEE

Fig. 10 shows the flow diagram of EEE algorithm. The shared key is used to split the input data. If the split is 64-bit long, it is added with the subkey; otherwise flipping and crossover take place. Flipping and crossover reduce the overhead of rotation used in RC5 encryption.

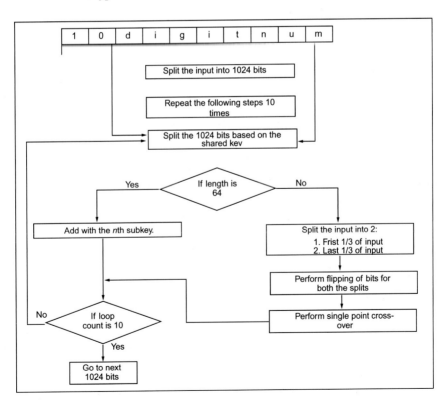

FIG. 10

Flow diagram of EEE.

Suppose six is present in the nth position of the shared key and then add it with the nth subkey. For example, if the shared key has six in fifth position then add that split with the fifth subkey and if six is present in ninth or 10th position add with first and second subkey, respectively.

3.9 Energy efficient decryption

Decryption is the exact reverse process of the encryption. The data received from the sender are split based on the shared key, and then further split is made if the length is not 64 and succeeded by single-point crossover and flipping of bits. If the length is 64 bits, then it is subtracted with the respective subkey.

4 Performance evaluation

4.1 Time complexity analysis

Time taken for the subkey generation, MAC generation, and encryption plays an important in energy conservation since time is directly proportional to the energy consumed, the processing time increases, and the energy consumed also increases.

Fig. 11 shows the comparison between RC5 and EESG for the time required for the subkey generation. X-axis is the number of subkeys that is processed. Y-axis represents the time taken to generate subkeys in millisecond (msec). It is seen that EESG takes approximately 76.2% lesser time than RC5 for a subkey generation.

Fig. 12 shows the comparison between SHA and LMGF for the time required for MAC generation. X-axis is the data size that is processed, represented in megabits (MB). Y-axis represents the time taken in millisecond (msec). It is seen that LMGF takes 62.7% lesser time than SHA for generating MAC.

Fig. 13 compares the time consumed during the encryption. X-axis is the data size that is processed, represented in megabits (MB). Y-axis represents the time consumed in milliseconds (msec). It is seen that EEE takes 63.1% lesser time than RC5 for encryption.

4.2 Analysis based on energy

Fig. 14 compares the energy consumed during the overall MAC generation. The power required for each component is considered individually, and the overall power consumed is calculated. The power consumed for LMGF is approximately 69% lesser when compared with SHA.

Fig. 15 shows the battery power remaining after the MAC generation. Initially the node is set with a full battery (100%). X-axis represents the remaining battery left in the node represented in percentage (%). X-axis represents the time taken in millisecond (msec). Y-axis represents the time taken in millisecond (msec). Y-axis is the data size that is processed, represented in megabits (MB). As the nodes process more data, their battery powers decrease. It is seen that the battery power remaining in LMGF is higher compared with SHA.

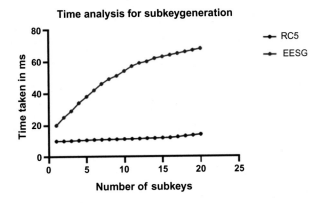

FIG. 11

Time comparison for subkey generation.

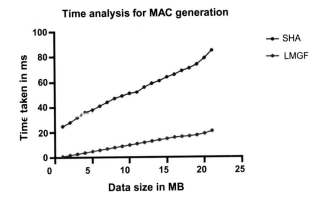

FIG. 12

Time taken for generating MAC.

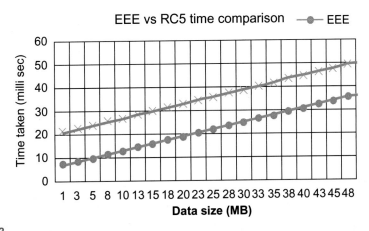

FIG. 13

EEE versus RC5 time comparison.

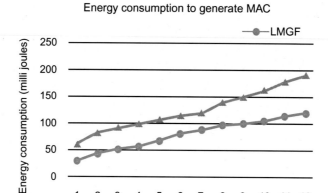

FIG. 14

Energy consumed to generate MAC.

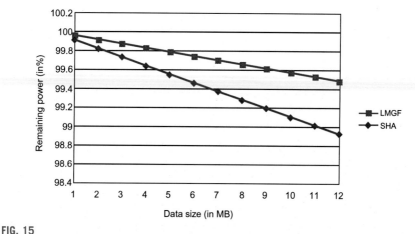

FIG. 15

Battery life of nodes running LMGF and SHA.

4.3 Security attacks

In general the MAC generated can be easily broken if the master key or subkeys generated are decoded. If the keys are known, then the input string can be changed, and the equivalent MAC will be sent to the destination. In such cases it is difficult to find any intruders in the system. The decoding can be done by analyzing the MAC generated. For example, a → 5, b → 6in such a case, it is easy to decode a string say abc. Any key can be broken easily, but the time taken to break the key is the essential factor. Here, two severe attacks, (i) the brute force attack and (ii) cryptanalysis attacks, are considered.

4.3.1 Brute force attack

Brute force attacks are not dependent on a specific algorithm, but depend only on the bit length of the text. So, to crack a 128-bit of data, the worst case is to try 2^{128} times. The time required at 1 decryption/μs is 5.4×10^{24} years, whereas the time required at 10^6 decryptions/μs is 5.4×10^{18} yrs. This is same for RC5, SHA, and EESG.

4.3.2 Cryptanalysis

Cryptanalysis seeks to exploit the weaknesses in a cryptographic algorithm. These attacks are based on hash functions and manipulate some properties of the algorithm and carry out some other attacks other than just exhaustive search. The resistance of a hash function to cryptanalysis is measured by comparing the latter's strength to the effort required for a brute force attack. The final sequence is obtained by analyzing the data to exploit and decode the entire functionality.

5 Conclusion

For any low-power device, the battery life, time, and energy of the network are limited. Here, it is shown that in EESG and LMGF, which includes the basic operations of XOR and shift is suitable to overcome the limitations of low-power devices. The result indicates that the algorithm achieves a reduction in hardware resources providing an energy-efficient network, thereby increasing their life span in the network.

EEE uses the concepts of genetic algorithm to provide lightweight encryption and uses subkeys to increase security. Experimental results prove that EEE is better than traditional RC5 algorithm.

Further reading

M. Ammar, G. Russello, B. Crispo, Internet of things: a survey on the security of IOT frameworks, J. Inform. Secur. Appl. 38 (2019) 8–27.

D. Dinu, Y. Le Corre, D. Khovratovich, L. Perrin, Triathlon of lightweight block ciphers for the Internet of things, J. Cryptogr. Eng. 9 (3) (2019) 283–302.

S.S. Kumar, S.A. Sherfin, A cryptographic encryption technique byte-spiral rotation encryption algorithm, J. Discret. Math. Sci. Cryptogr. 22 (3) (2019).

K.V. Pradeep, V. Vijayakumar, V. Subramaniaswamy, An efficient framework for sharing a file in a secure manner using asymmetric key distribution management in cloud environment. J. Comput. Netw. Commun. 2019 (2019) 9852472, https://doi.org/10.1155/2019/9852472.

S.L. Keoh, S.S. Kumar, H. Tschofenig, Securing the internet of things: a standardization perspective, IEEE Internet Things J. 1 (3) (2014) 265–275.

M.P. Grabovica, S. Pezer, D and Knežević, V., Provided security measures of enabling technologies in internet of things (IoT): a survey, in: Zooming Innovation in Consumer Electronics International Conference, 2016, pp. 28–31.

C. Pfister, Getting Started With the Internet of Things, Make: Community, Santa Rosa, CA, 2011.

William Stallings, n.d. Cryptography and Network Security, ISBN 978-0-13-606741-2, Page no 10, 19, 367.

F. Li, P. Xiong, Practical secure communication for integrating wireless sensor networks into the internet of things, IEEE Sensors J. 13 (10) (2013) 3677–3684.

M. Wei, Z. Wang, K. Kim, An implementation of security mechanism in chip for industrial wireless network, in: Information Networking Conference, 2017, pp. 265–270.

C. Hennebert, J.D. Santos, Security protocols and privacy issues into 6Lowpan stack: a synthesis, IEEE Internet Things J. 1 (5) (2014) 384–398.

J. Granjal, E. Monteiro, J.S. Silva, Security for the internet of things: a survey of existing protocols and open research issues, IEEE Commun. Surv. Tutorials 17 (3) (2015) 1294–1312.

A.A. Fuqaha, M. Guizani, M. Mohammadi, M. Aledhari, M. Ayyash, Internet of things: a survey on enabling technologies, protocols, and applications, IEEE Commun. Surv. Tutorials 17 (4) (2015) 2347–2376.

S. Sharma, R.K. Bansal, S. Bansal, Issues and challenges in wireless sensor network, in: International Conference on Machine Intelligence Research and Advancement, 2013, pp. 58–62.

J.L. Hernandez-Ramos, J.B. Bernabe, A.F. Skarmeta, Managing context information for adaptive security in IoT environments, in: 29th International Conference on Advanced Information Networking and Applications Workshops, 2015, pp. 676–681.

W.K. Koo, H. Lee, Y.H. Kim, D.H. Lee, Implementation and analysis of new lightweight cryptographic algorithm suitable for wireless sensor networks, in: International Conference on Information Security and Assurance, 2008, pp. 73–76.

T.B. Singha, L. Jain, B. Kant, Negligible time-consuming RC5 remote decoding technique, in: International Conference on Computer Communication and Informatics, 2014.

U. Jasmin, R. Velayutham, Enhancing the security in signature verification for WSN with cryptographic, in: International Conference on Circuit, Power and Computing Technologies, 2014, pp. 1584–1588.

C. Sarkar, A.U. Nambi, R.V. Prasad, A. Rahim, R. Neisse, G. Baldini, DIAT: a scalable distributed architecture for IoT, IEEE Internet Things J. 2 (3) (2015) 230–239.

L. Catarinucci, D.D. Donno, L. Mainetti, L. Palano, L. Patrono, M.L. Stefanizzi, L. Tarricone, An IoT-aware architecture for smart healthcare systems, IEEE Internet Things J. 2 (6) (2015) 515–526.

S. Hemalatha, V. Rajamani, VMIS: an improved security mechanism for WSN applications, in: International Conference on Science Engineering and Management Research, 2014, pp. 1–3.

R. Fantacci, T. Pecorella, R. Viti, C. Carlini, A network architecture solution for efficient IoT Wsn backhauling: challenges and opportunities, IEEE Wirel. Commun. 21 (4) (2014) 113–119.

J. Singh, B. Kumar, A. Khatri, Improving stored data security in cloud using Rc5 algorithm, in: Nirma University International Conference, 2012, pp. 1–5.

H.J. Kim, H.S. Chang, J.J. Suh, T.S. Shon, A study on device security in IoT convergence, in: Industrial Engineering, Management Science and Application International Conference, 2016, pp. 1–4.

V. Sai, M.H. Mickle, Exploring energy efficient architectures in passive wireless nodes for IoT applications, IEEE Circuits Syst. Mag. 14 (2) (2014) 48–54.

K.L. Lueth, IoT basics: getting started with the internet of things, March, 2015, Available from: http://iot-analytics.com/wp/wp-content/uploads/2015/03/2015-March-Whitepaper-IoT-basics-Getting-started-with-the-Internet-of-Things.pdf.

GSMA Technology, *Understanding the Internet of Things (IoT)*, 2014, Available from: https://www.gsma.com/iot/wp-content/uploads/2014/08/cl_iot_wp_07_14.pdf.

D. Ganesan, A. Cerpa, Y. Yu, Z. Jerry, D. Estrin, Networking issues in wireless sensor networks, in: Computing and Communication in Distributed Sensor Networks Conference, 64 (7) 2004, pp. 799–814.

S. Sukhwinder, Issues and challenges in wireless sensor networks, in: International Conference on Machine Intelligence and Research Advancement, 2013, pp. 58–62.

B. Bhushan, G. Sahoo, Recent advances in attacks, technical challenges, vulnerabilities and their countermeasures in wireless sensor networks. Wirel. Pers. Commun. 98 (2) (2017) 2037–2077, https://doi.org/10.1007/s11277-017-4962-0.

B. Bhushan, G. Sahoo, ISFC-BLS (intelligent and secured fuzzy clustering algorithm using balanced load sub-cluster formation) in WSN environment. Wirel. Pers. Commun. (2019), https://doi.org/10.1007/s11277-019-06948-0.

P. Sinha, V.K. Jha, A.K. Rai, B. Bhushan, Security vulnerabilities, attacks and countermeasures in wireless sensor networks at various layers of OSI reference model: a survey. in: 2017 International Conference on Signal Processing and Communication (ICSPC), 2017, https://doi.org/10.1109/cspc.2017.8305855.

S. Jaitly, H. Malhotra, B. Bhushan, Security vulnerabilities and countermeasures against jamming attacks in Wireless Sensor Networks: a survey. in: 2017 International Conference on Computer, Communications and Electronics (Comptelix), 2017, https://doi.org/10.1109/comptelix.2017.8004033.

Blockchain as a solution for security attacks in named data networking of things

10

Sukriti Goyal[a], Nikhil Sharma[a], Ila Kaushik[b], and Bharat Bhushan[c]

[a]*HMR Institute of Technology & Management, Delhi, India* [b]*Krishna Institute of Engineering & Technology, Ghaziabad, Uttar Pradesh, India* [c]*Birla Institute of Technology, Ranchi, India*

Chapter outline

1 Introduction

The IoT stands for Internet of Things, which is envisaged in the engineering of the Internet to interrelate/interdepend components of over the whole earth by transmittable gadgets/devices for understanding/sensing, processing of data, and

211

Security and privacy issues in IoT devices and sensor networks. https://doi.org/10.1016/B978-0-12-821255-4.00010-9

communication with data. Internet of Things (IoT) structure has been extensively used in various fields that are smart cities, homes, farming and grids, digital health, sensing of environment, and planning for urban area, and all of this just because of the inventiveness and intelligence of the embedded stuff. Still, constructing IoT with all physical stuff in a combined form with Internet faces two disputes [1]:

(i) Process of interaction of IoT gadgets both globally and locally.
(ii) And security or privacy of the communicated data that is linked to the network.

With the advancement of Information Technology, the IoT has been unified into our existence. The Internet of Things is spreading at a fast rate, and it is predicted by some reports that around by 2020 the Internet of Things gadgets will increase to 25.8 billion, which are 30 times more than the predicted count of gadgets redistributed in 2009, and is also far more than 7.4 billion tablets, PCs, and smartphones that all are requisite to be in work by 2020. It is essential to create an Internet of Things stack, systematize protocols, and build the complete and suitable layers for the structure that will serve facilities to gadgets of Internet of Things to reach at such a giant growth. Presently, most of the IoT clarifications depend on the concentrated client-server prototype, which through the Internet are connected to cloud servers.

The Internet of Things is a huge network of related gadgets and people, all of which gather and transmit data about the surrounding around them and about the manner they are used. Considering the growth of IoT in human's life, one thing is completely clear that "Connected world is the future," The Internet of Things gives an assurance of innovative or revolutionary solutions in every field either it is the topic of smart agriculture or of smart city.

Current Internet of Things groundworks is explained to clarify the connectedness of substantial things. Out of groundworks, most of these implement the connectivity issues by the ultimate predominant Internet Protocol (IP) in the host-to-host manner. In this network, which is host based, it requires both a problem and long mappings of junction addresses. Moreover, host-to-host transmission is not appropriate to IoT, because in the passage of network connection, Internet Protocol facilitated system of connections is disrupted. Currently, to resolve the irruptions (attacks) of IoT, many firms propose to apply NDN [2], stands for Named Data Networking by claiming the benefit of paradigm of Named Data including both data and host that are titled by the naming projects, with the distribution of content, which is swapped to content adapted transmission from host-to-host manner. In such way the network distributes the named (titled) data and send further straightly on titles carrying operation definitions.

In the previous research and surveys, many types of security or privacy irruptions and explanations were interrogated in the Named Data Networking framework. Cache pollution, unauthorized access, resource exhaustion, dismissal of services, path infiltration, etc. [3] all these are contained by the usual security irruptions. It is deplorable that security explanations practiced to Named Data Networking (NDN) cannot straightly acclimate to Named Data Networking of Things (NDN of Things) as the irregular, dynamic, and assorted connectivity of things, in spite that explanations were carried out to contend up with the aforementioned irruptions in the prospect of caching, naming (titling), and routing.

This article gives a survey of security irruptions in Named Data Networking of Things, to enable readers to recognize the dangers rooted by the surveillance irruptions or security problems and embolden to interrogate the best result for Named Data Networking of Things. Particularly, this chapter is illustrating the analysis of NDN of Things. After that, security irruptions in regard to information or data regeneration, with an overview of caching and data distribution. Then, it is providing current solutions for security irruptions in NDN of Things with an explanation enabled with blockchain [4] to solve security irruptions in reference of NDN of Things and also the basics and architecture of distributed blockchain for Named Data Networking of Things with the key features of its design are presented. In addition, data sharing protocol, AES algorithm with its features, and the forms of blockchain (that are basically classified into four categories) are discussed. Furthermore a detailed study is introduced on how to accommodate the blockchain technology to the particular requirements of Internet of Things to create BIoT applications, that is, Internet of Things applications based on the technology of blockchain. Moreover the need for using the blockchain and what are the most appropriate applications of BIoT are detailed. Then, existing issues of BIoT applications in terms of security, privacy, etc. are described that affect the growth, architecture, and optimization of an application of BIoT. At last, some further issues and recommendations are illustrated with the motive of directing the oncoming developers and investigators of BIoT on some of the challenges that will have to be unraveled prior to the deployment of the coming generation applications of blockchain-based Internet of Things [5].

2 Security attacks in NDN of things (NDNoT)

There are many security concerns that need to be addressed. As NDN of Things is inherited from Named Data Networking groundwork, all the irruptions of Named Data Networking also come in sight in NDN of Things, such as cache pollution, unauthorized access, resource exhaustion, dismissal of services, and path infiltration. Out of list of security irruptions, numerous of security irruptions in Named Data Networking have been discussed in many past operations. While in the current article, the security irruptions of the framework complimentary network are primarily explained. Particularly, typical irruptions are interrogated in the form of cache misappropriation, interest flooding, selfish irruption, data phishing, and miscellaneous irruptions too. Let's understand these attacks one by one.

2.1 Cache misappropriation

Security of Named Data Networking hidden supplies is split into cache poisoning and cache pollution. To employ the cache repository in a device, the assailant employs the fake content that results in the extrusion of original content from the cache storage, and this process is named as "Cache Pollution." In the same manner, assailant occupies the Data Packet with the fictitious payload to occupy the cache storage, which results in satisfaction of Interest Packet by the fraudulent content packet, and draws the illegitimate occupation of cache repository, referred as "cache poisoning." Both "cache pollution" and "cache poisoning" are coming under to the cache misappropriation; at the same time, both have a similar intention of using fake or illegal data to fill the cache storage of a device.

Although many firms and researches tried to explain the irruption of cache misappropriation in Named Data Networking and gave many ideas. An approach is presented by cache shield [6] to decrease "cache pollution" by identifying data with their demand and to resolve the problem of "cache poisoning" that can be resolved by authentication of each and every packet. But it introduces verification in huge amount, which cannot be possible to apply in the Internet of Things in those worthy gadgets that have defined computational scope [7, 8]. On one side, as the Interest Packet and Data Packet treat the system of connections through distinct-distinct paths, but the reputation of data is challenging to be investigated. While, on another side, the batch authentication of data on the cache requires the dissemination of keys in decentralized Internet of Things network, which is also a difficult task.

2.2 Interest flooding

As we have seen in the preceding text, NDN of Things includes hosts who transmit Interest Packets as the typical action to appeal for the required and wanted data. Still, sometimes malicious consumers may transmit illegal or not worthy or fake Interest Packets in the system to block system sources, for example, power, bandwidth, and storage.

There are some current solutions [9], which can margin the counting of Interest Packets from the end hosts, which are at the end to neglect the Interest flooding. But the approachability cannot be effortlessly engaged to Named Data Networking of Things (NDN of Things). Firstly, it will be very challenging for system of connections to determine the holders of the packets and as we know Interest Packets are data oriented in NDN of Things. But here are some methods [10] that can margin the counting of Interest Packets from the host by putting the signature of owner of Interest packet as a component of Interest Packet to verify the holder of that packet. This approach can be occupied by concentrating Naming Data Networking. However, in the reference of Named Data Networking of Things, hosts are mostly mobile and dynamic in nature, and comprise a decentralized system of connections surrounding, in which the rate limit cannot be handled. Until the rate limit is not reached, an assailant may send Interest Packets to one definite host, but when the limit rate is reached, the assailant may stop sending the Interest Packets to that host. After that, it moves to another place and start sending Interest Packets to another host until the limit rate is reached. In this way the limit rate is rifted. Apart from this, there are some more methods that tried to control the flooding by rankings, but these rankings are also facing the problems of disseminated and dynamic features of Internet of Things.

2.3 Selfish irruption

Many harsh security irruptions of NDN of Things conclude, due to the self-indulgence of hosts. In the reference of Named Data Networking of Things, specifically, in the framework-free connection, hosts present as promoters of Interest or Data Packets. At the same time, those hosts may act self-indulgent and do not equip the promoting functions of the Data or Interest due to the circumspection of storage and computation of hosts of Internet of Things, concluding denial-of-service (DoS) irruptions. During the processing, when a packet is transmitted to the assailant, there is a probability of detecting the name of the packet of the assailants and filter out the name component that they do not want to equip and block the more leading. In worst case, due to self-indulgence behavior, an assailant may not transfer any more leading services for other ones.

To handle this kind of irruptions and to persuade packet promotion, each host of that packet requires the encouragement. In the past work an access of enticement for disseminating system has been evaluated in discrete schemes [11].

2.4 Data phishing

To reach to the data, Interest Packets are supervised to the data section where the wanted content is concentrated. By the prominent data names in the sector, the data section is recognized. After that, with the help of hosts who are more usual with the content that appealed by Interest Packet, the Interest Packet is routed. Then, this kind of manner inspires malignant hosts to plan phishing intrusion on Data Packets. Particularly, to encourage an Interest Packet, an assailant will prepare a lot of fraud content and transmit fraud and nonworthy data names in a huge amount that contain may be the same or associated constituents of names of Interest Packet. Now, after dealing with the one host, when the assailant faces the other hosts, it starts distributing the assemble names of data with those hosts, to hide the recognition of their data section with the help of those distributed and assembled data names. Sometimes what happens when an Interest Packet appeals these data and gets fascinated to the data sector, but cannot get it wanted and required data with gratification. In data phishing irruption, basically an assailant can fascinate the Interest Packets but refuse to supply the data and leads to security irruption.

In current practices, to verify data from each and every correspondent contact, Interest Packet is endorsed by the data owner itself. In spite that, by these practices with the help of signatures, the Interest Packets can be authenticated. Now, in Named Data Networking of Things, Interest Packet verification is facing new types of disputes.

- On one side, Interest Packet takes computing power of hosts in huge amount for authentication, which is impracticable for the Internet of Things (IoT) network.
- And, on another side, framing Interest sector requires only the names of data rather than requiring the complete Interest Packet.

The assembled data names by assailants cannot be neglected by the verification access.

2.5 Miscellaneous irruptions

Various miscellaneous irruptions that appear on Named Data Networking also come into existence in Named Data networking of Things. Illegitimate access, playback irruption, and packet detecting are some examples of miscellaneous irruptions.

Playback irruption and packet detection assign to those irruptions in which an assailant tries to approach the packet of data malignantly and attempt to detect the data of packet and finally, change, pause, or block the packet for more forwarding. After that the assailant approaches the packet and changes the payload of the packet in such a way so that the wanted content cannot collect by the committed hosts. Moreover, to misuse the bandwidth of the system's network, the assailant also replies the Interest Packet with many times. In addition, an assailant can also plan a replay irruption that customizes the Interest Packet so that the assailant can pause wanted data of the interest appealer and completely misuse the resources of a network.

Out of many, here is one miscellaneous irruption that is illegitimate access in which without any authorization an assailant is able to access the data transmitted to a particular host and can also use any replica of the packet for an illegitimate approach that results in difficulty to control the access in Named Data Networking of Things. By using packets encryption, current conclusions attempt to resolve these challenges. But at some place, these conclusions meet the problems of key dissemination and management of key, specifically in disseminated Named Data Networking of Things (NDN of Things) [12, 13].

3 Blockchain in NDNoT

All these described security irruptions cannot be neglected by current security results due to dynamic features and fragmentation of Named Data Networking of Things. So, to handle the aforementioned irruptions of NDN of Things, an approach of blockchain is conferred.

3.1 Blockchain basics

Blockchain technology is neither an app, nor a company, but a completely new way of data documentation on the Internet. In the year of 1991, blockchain came to existence. For safety of digital scripts, an association of analysts wanted to develop a technology of which, all that documents neither misdated nor modified in the future for any illegal purpose. So, in the simple and easy terms, the blockchain is a time-stamped series of unmodifiable records of the data that are linked by cryptographic hashes and maintained by group of systems and not owned by any single entity. The network of blockchain has no any centralized authority. In other terms, blockchain is a peer-to-peer disseminated ledger that is forged by consensus and linked to a system for smart contracts and other dependable technologies. Similar to distributed ledger whose data are transmitted among the peers' system blockchain has the same functioning. Distributed Ledger Technology, abbreviated as DLT, is a classification of database technology that contains either characteristics of blockchain technology or itself blockchain technology.

Now, why it is known as "blockchain"? The answer is that blockchain owes its name basically to how it does function and the way in which it stores all the data, especially that information which is gathered in the form of blocks, which are linked in a manner to form a chain with the other blocks of the same information. So, it is the process of linking the blocks into a chain of similar information that makes the information trustworthy stored on that blockchain. Once the whole data are documented in a block of chain, then it cannot be modified or changed without having modification in each and every block that occurs after that block.

The objective behind the blockchain technology is actually innovative and forthright, being earnestly explored by lakhs of people from the whole world who participate in many projects and that participation push the technology of the blockchain industry forward.

3.2 Blockchain architecture

In accordance with distinct needs, distinct blockchain structures are architecture. Decentralization is property of this structure. After the loading of all the blocks/sections from the starting one to the last one by the nodes, initialization is achieved.

For disseminated networks [14], blockchain acts as a security approach that also provides clarity, confidence, and information security to all the digital disseminated scripts. In the financial industries, blockchain design is widely used. Apart from this, recently, this technology hired not only in cryptocurrencies but also in maintaining records, digital signatory, and smart contract. It also supplies as a database of community that registers all actions in the structure, which handles a string of hashed sections and guarded by cryptography. In the process, firstly, in the blockchain infrastructure, each and every host enters its identification and services. After this step, in the blockchain, each and every activity is registered by a trustworthy host and then, on the network, the blockchain is publicized to all the hosts. Using this benefit, blockchain-dependent identification and access management systems can reinforce the security of Internet of Things. But, at some point, influence of blockchain to NDN of Things is bearing new disputes, that are as follows:

(i) In blockchain, what data or info requires to be managed?
(ii) How to manage records of Data or Interest?
(iii) How to resolve the problem of irregularly linked hosts?

That's why, to state mentioned disputes, the Named Data Networking Blockchain of Things is constructed. An unrevealed and difficult to summarize key is held by each and every host that is recorded in its own repository. In a hashed way the Data's or Interest's path is managed by the blockchain technology. An example of a disseminated blockchain of Named Data Networking of Things is shown in Fig. 1, which contains an array of sections/blocks and that each section consists a hashed transit set. Excluding the starting section, remaining all blocks holds a head pointer that is connected to a preceding section. Also, each block contains the timestamp that reports that time when the section is written.

To manage blockchain in Named Data Networking of Things, by the approach of proof of work (PoW) [15], design-authorized hosts act as the "miners" to try to note down the blocks or sections into the blockchain. For example, that host will be preferred to compose the blocks into the blockchain, who will be able to solve out the puzzle(s) of hash first.

In NDN of Things the smart contract of the blockchain performs in the following way: in smart contract, when a host1 transmits a Data or Interest Packet by straightforward network, to a host who is in contact with it, then the nearby hosts of host1 store the signed and hashed data transformation, from the exporter to the receiver and also the equivalent timestamp. Then, after meeting with the design-authorized hosts, these nearby hosts receive a renovated blockchain from the host and transmit the transformed data/info to the host at the same moment, and after receiving that the host circulates that information in the design-authorized network and a transition block or section is constituted. To acquire the chance of composing, the transition block in blockchain, the proof of work is initialized by each host. For the purpose of inspiring the hosts for blockchain composing, those hosts who are elected are awarded with excessive rate of Data or Interest.

As displayed in Fig. 2, on the behalf of model of common network of blockchain, an abstract model of six layer is presented to illustrate the design of the blockchain for Named Data Networking of Things at the level of abstraction. Now, let's understand the main characteristics of the introduced model in brief one by one;

A. *Physical layer*

This layer contains numerous kinds of objects, which can be present in automobiles moving on roadways, domestic detectors and in digital resources having illusive

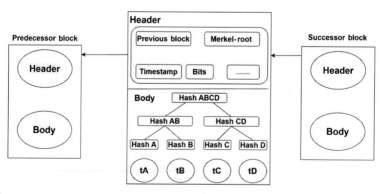

FIG. 1

The disseminated blockchain for NDN of things.

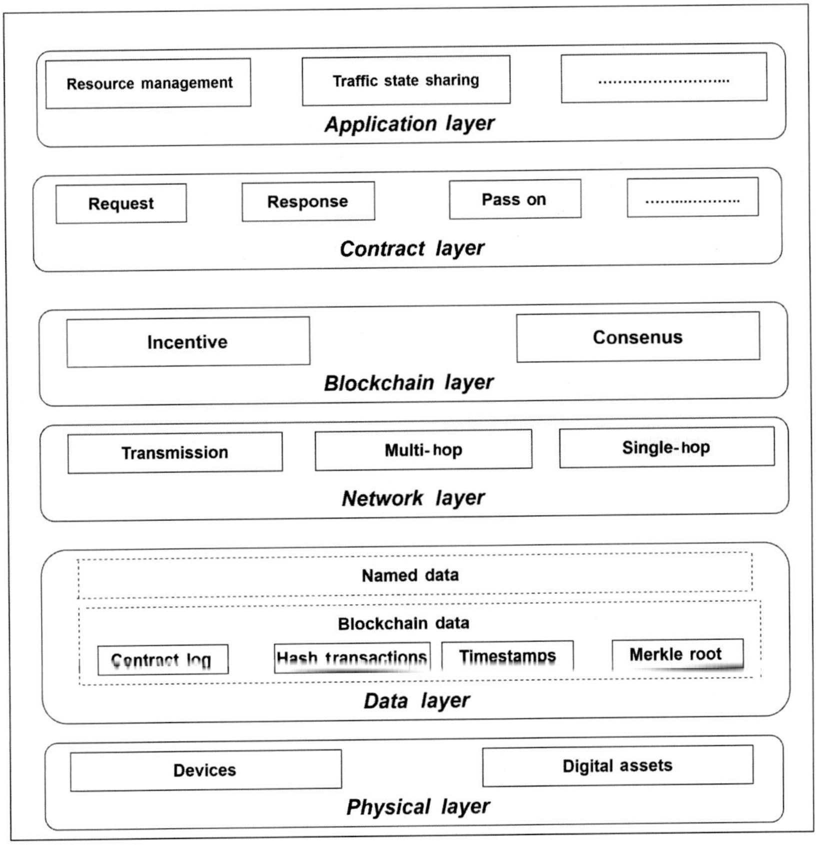

FIG. 2

The architecture of NDN blockchain of things.

nodes. All of these objects are formerly maintained by the medium of both the concentrated system and the safe and trustworthy connection of the blockchain.

B. *Data link layer*

The layer of data is comprised of two segments of data. The first segment is called as "Data" in NDN of Things and the second one, is a component in the blockchain as a data, containing contract logs, timestamps, and hash of transactions.

C. *Network layer*

The layer of the network presents benefits having common functioning that can be easily utilized by the layer of blockchain for synchronization or by layer of application to propagate. This layer also gives fundamental network communication abilities for broadcast, multihop and single-hop for the application, contract, and blockchain layer.

D. *Blockchain layer*

The two important processes of the blockchain are contained by the layer of blockchain.

- First one is consensus, which assures that the common lengthy string in all the systems inside the blockchain architecture is contributed by the complete net.
- Second one is the incentive awards for those particular nodes, which carry out the blockchain authentication operation before the addition of any entity in the design of blockchain.

E. *Contract layer*

All the conclusions and the implementation of the actions of the computer inside the architecture of blockchain (or nodes) in the NDN of Things are considered by this layer. It also fulfills the process of requesting, replying and forwarding from the nodes. At the same moment the smart contract works as tamper proof.

F. *Application layer*

The application layer in the Named Data Networking of Things contains numerous operations, for example, digital data maintenance and traffic situations.

3.3 Key features of design of blockchain

An architecture of blockchain acquires a various advantage for business purposes. Numerous entrenched features are as follows:

- *Anonymity:* Each and every blockchain system participator has a created address rather than the user's identification. Due to this, it secures the user's identification anonymous, especially in that blockchain design, which is public.
- *Cryptography:* Due to the cryptographic evidence and difficult computations among the parties that are involved, all the contracts of blockchain are secured and trusted.
- *Clarity:* There is a very less probability of blockchain to be fraudulent, as it is very complex and needs a very large power of computation to rewrite the entire blockchain network.
- *Perpetuation:* It is the feature of blockchain architecture due to which any records that are recorded in the blockchain cannot be modified or edited or removed from the database.
- *Decentralization:* In the design of blockchain, each party can access the entire disseminated database in the architecture.

4 Security investigation of NDN blockchain of things

The security investigation of Named Data Networking Blockchain of Things is proposed to prohibit security irruptions. The NDNBoT maintains all the actions of demanding, responding, and sending further and the Data or Interest hash by the smart contract. Also, there is no chance of leakage of user's secrecy as both the user and data identifiers are anonymous. In addition, it has the feature of immutability in which all the records cannot be edited or removed in future because of the architecture of the blockchain of the hashed series of blocks in a specialized arrangement. It is observed that the blockchain having smart contract is safer and more trustworthy and the model of Named Data Networking Blockchain of Things, which is able to identify and locate the assailants. Here the circumstances under these four kinds of irruptions (which are previously proposed in a preceding sections) are analyzed.

The attacks are as follows:

- Interest flooding
- Data phishing
- Cache misappropriation
- Self-indulgence

Let's discuss the circumstances under these irruptions one by one.

a. *Interest flooding*

When a malignant host transmits Interest Packets in a very large amount, sometimes the number of packets surpasses the limit of rate, which is identified by the other members by interrogating the list of blockchain. Then the smart contract of blockchain avoids the interest flooding irruption rather than to put more efforts on mining.

b. *Data phishing*

In case of data phishing irruption, interchanging the titles of packets of data thoroughly depends on blockchain. In the blockchain, each and every hash of the info is taped. If in any case, packets of info are not permeated with it, and then the host will have no justification to create fraud titles/names. Or else, they can be effortlessly inspected by the other ones.

c. *Cache misappropriation*

From previous discussions, it is clear that the misappropriation of cache occurs due to the placement of unauthorized data. By the smart contract the source and the appeal of the data can be copied either by event(s) log or variable of state, when a host tries to employ fraud or unfamiliar data.

d. *Self-indulgence*

As we discussed in preceding sections, the self-indulgence of hosts is one of the major issues of Internet of Things. The irruption self-indulgence undoubtedly is also a problem for hosts to compose sections into the blockchain. For the purpose of inspiring the hosts to compose for blockchain, those hosts who are elected are awarded with excessive rate of Data or Interest, after which they get the powers of transmitting many more Interest Packets to the Named Data Networking of Things.

Also a number of miscellaneous irruptions can be solved by the solutions of blockchain, which will provide the most efficient outcomes.

An Named Data Network (NDN) can be implemented by the user that depends on an organized chain to assess the usefulness of the Named Data Networking model, which is based on the blockchain. Data Distributing Protocol, Data Approach Regulation, and Event Log of Smart Contract, all these three are included in the technological details of it.

In the preceding phase mentioned model, the smart contract is answerable for the functional transcripts and data approach regulation. Granting and revoking the permission of its susceptive content can be done by the data/content holder with granularity. All these can be possible to happen with the help of functions in the smart contract. Those functions can only be approached by the owner of the contact. Let's understand those functions one by one:

(i) *Permission/authorization*

In this function, one's address of Ethereum account and public key can be transmitted by the owner to smart contract in order that that account in the form of a permissible suppliant of the holder's content will be considered by the smart contract.

(ii) *Revocation*

In this, one's address of Ethereum account can be passed to the smart contract by the owner(s) in order that that account holder will not anymore be treated as the permissible consumer by the smart contract.

An event log will be triggered by the two activities, which are present in the smart contract, when authorized consumers transmit transactions to call those activities. Inside that transaction, in the Ethereum blockchain, that particular event log will be unchangeably stocked.

- *Data request*

As presented in Fig. 3A, to get the decryption key of data, a transaction can be transmitted by the requester/appealer of the data to the smart contract of the owner of the data. The contract will enter that appeal and procreate an event of appeal/request, in the case, if the appealer is authorized. It holds the type of the log as the "Request," requester's/appealer's address, appealed data's hash, appealer's public key, and also a timestamp.

- *Data receive/acceptance*

As presented in Fig. 3B, transaction can be transmitted by the owner of the content to the smart contract of content appealer to transmit the decrypting key of data.

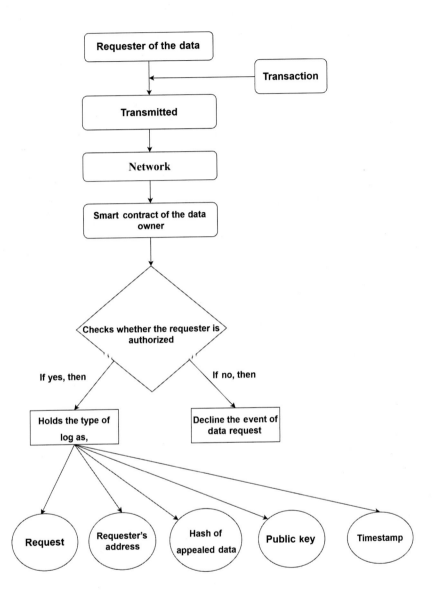

Fig. 3 See figure legend on next page.

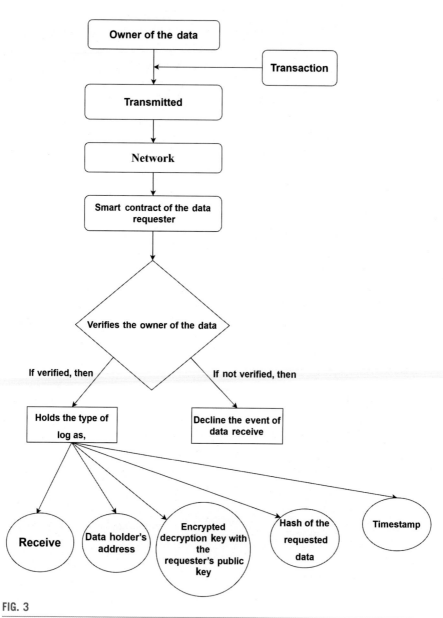

FIG. 3

Data structure of contract event log: (A) data request and (B) data receive.

After the verification of the content owner that received activity will be entered into the smart contract and then procreate an event of acceptance/receive. A type of log as "Receive," content holder's address, decryption key encrypted with the appealer's/requester's public key, a timestamp, and the hash of the appealed/requested data will be hold by it.

In case of broadcasting a confidential data, it requires to be encrypted in the form of cipher text. A data transmitting protocol with access practices based on the blockchain is projected to assure the security of transmitting of data. The preceding mentioned protocol not only regulates the approach of Named Data (ND) with the access practices in the blockchain but also assures that all the performed transmitting activities have held their equivalent unchangeable log(s) for both clarity and verification. Each smart contract manages a permissible consumers' list to recognize the authorized status of appealers/requesters. That list constitutes the users' address of Ethereum account and their public key. The process of data sharing protocol is shown in Fig. 4 and split into the steps that are as follows.

- *Step 1:* Firstly a named data holder updates his/her smart contract by adding updated data/info hash in it, with the aim of transmitting it to his/her trustworthy "friend(s)."
- *Step 2:* After that, owner's encrypted content, data's lifetime, timestamp, address of Ethereum account, address of smart contract, and hash data are broadcasted by the owner of Named Data itself. Leakage of plaintext to the unauthorized node(s) is prevented by the encrypted data. The usage of info hash in the process is to

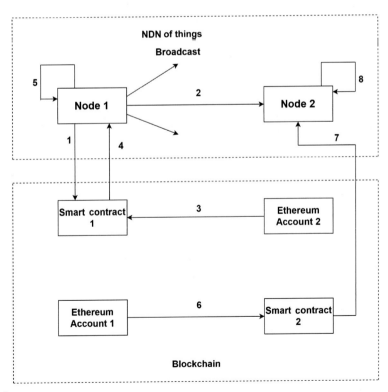

FIG. 4

Data sharing protocol.

transmit the transaction that is appealed to the content owner's smart contract. The adequate duration of that content is illustrated by both the lifetime and a timestamp.

- *Step 3:* Then, after getting the transmitted content, a transaction with the hash of data is sent by the Named Data appealer so that the decrypted key of data appeals/ requests to smart contract of the data owner.
- *Step 4:* In the smart contract of the content owner, the appealing function replies to transaction request and authenticates the identity of appealer and the clarity of the appealed data. In case, if the verification is cleared, an event of a request will be procreated by the smart contract.
- *Step 5:* Then, that request/appeal event is observed by the owner of the data, and after that the decrypted key with the appealer's public key is encrypted by the data owner and a transaction with the encrypted key is transmitted to the smart contract of the requester of the data.
- *Step 6:* The data holder's identification takes place by the appealing function of smart contract of data appealer and the observation of the procreation of the request event takes place by the appealer of the data.
- *Step 7:* After the observation of appealed event, decoding of the encrypted key can be done by the requester of the data with both the encrypted data with the content decoded key and his/her private key.

Each account of the Ethereum contains a couple of keys to boost digital signature and encrypted transmission together with another duo of keys that are private and public key, which is used for encrypted decrypted key communication. Advanced Encryption Standard 128 (AES128) algorithm is used for each and every user or nodes to encrypt its private important content or info or data. So, the size of the decoded key is 128 bits.

A network of Internet of Things is generated that depends on Ethereum to figure out the procedure of both mining and recording into the blockchain. That network comprises three in count Internet of Things systems inside the architecture of blockchain or users and each user or system has the ability, that is,

- first node has the ability to perform mining,
- second one has the ability to perform smart contract,
- The last one, third node has the ability to perform transmission.

The complete system executes inside the virtual device/gadget together with the identical implements as of IoT gadgets, so the Internet of Things gadget can be simulated by just modifying the CPU and memory limits. Only one virtual gadget is used to represent an IoT user or computer within the architecture of blockchain. Then, we figure out what is the cost of CPU and occupation of memory in the in the circumstances of 3.3GHz one-core CPU having 1G memory and also, 3.3GHz dual-core CPU having 2G memory. Proceeding further, because of the reason that the process of mining occupies the most important role for utilization of Internet of Things gadget and speed of computation, only the evaluation of the accomplishment of mining on the Internet of Things gadget takes place.

4.1 AES128 algorithm

Advanced Encryption Standard is abbreviated as AES, which is a process of encrypting a content (results in cipher content), which includes both a cryptographic algorithm and key chosen by the government of the United States to secure the confidential data and is carried out in both hardware and software in the entire world to encrypt the confidential info or data (Fig. 5).

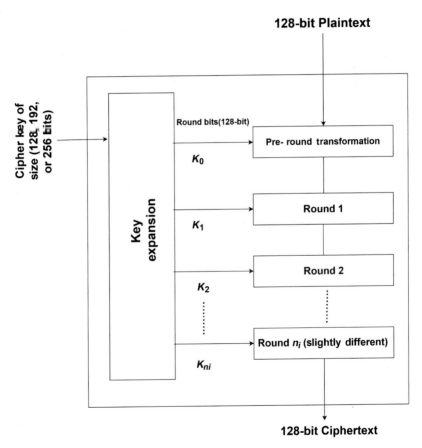

FIG. 5

Structure of AES (Advanced Encryption Standard algorithm).

Table 1 The relationship between the size of Cipher key and number of rounds and size.

Size of key	n_i
128	10
192	12
256	14

The characteristics of the Advanced Encryption Standard algorithm are as follows:

a. Software is implementable in both Java and C.
b. Much faster and powerful than the Triple-DES.
c. Having 128-bit data (128/192/256-bit keys) that's why also known as AES128.
d. Provides complete specification and design details.

Table 1 shows the relationship between the size of Cipher key and number of rounds and size.

Advanced encryption standard algorithm

1. Cipher *(byte in [Nb*4], byte out [Nb*4], word wd [Nb*(Nr + 1)])*
2. begin
3. byte state *[4, Nb]*
4. state = in
5. AddRoundKey *(state, wd [0, Nb-1])*
6. for round = *1 step 1 to Nr-1*
7. SubBytes*(state)*
8. ShiftRow*(state)*
9. MixColumn*(state)*
10. AddRoundKey*(state, wd[Nb*round, (round + 1)*Nb-1)*
11. *end for loop*
12. SubBytes*(state)*
13. ShiftRow*(state)*
14. AddRound Key (state, wd *[Nb*Nr, (Nr + 1) *Nb-1)*
15. out = *state*
16. End

---

5 Types of blockchain

Depending on the documented or recorded data, availability of that data and also what operations can be implemented by a user, there are different-different forms of blockchain, which are as follow (Table 2):

- Private blockchain
- Public blockchain
- Consortium blockchain
- Hybrid blockchain

Table 2 Types of blockchain.

Private blockchain	Public blockchain	Consortium blockchain	Hybrid blockchain
As its name suggests, private blockchain in other terms, is also known as permissioned blockchain. It is a private property of an organization or an individual	Public blockchain is transparent and an open source. It is a type of blockchain, which is basically— "by the people, for the people and of the people"	Consortium blockchain is a subclassification of the private blockchain and semiprivate in nature having all the same advantages of a private type blockchain and can be named as semidecentralized blockchain	Hybrid blockchain is the combination of advantages of both the public blockchain and the private blockchain
Participants need authorization to join the network(s)	In this case of blockchain, any user, developer, miner, or member can connect with the network of blockchain without the permission of any third party	In this case of blockchain, the network is governed by a group of organization rather than an individual or a single entity	By grasping the advantages of both public and private blockchain and optimizing the processes, any group of organizations can provide read-only access to their users and write access internally to providers so that they can improve the clarity of the processes in the blockchain
In this type of blockchain, all transactions are private and are only available to the surrounding participants that have been provided authorization to join the network	Public blockchain is designed in such a way that no entity or individual is in charge of which transactions are recorded and which one are processed, which means that there is no one in this blockchain to take decisions. Then the decisions will be taken by various procedures of decentralized consensus	Therefore this type of blockchain provides security that is inherited from the public blockchain as it is granted to a group of authorized entities	It is completely customizable because hybrid blockchain's authority can decide, which transactions should made public and who can participate in the network

Continued

Table 2 Types of blockchain—cont'd

Private blockchain	Public blockchain	Consortium blockchain	Hybrid blockchain
So, it can be seen that many private blockchains are also considered as permissioned to maintain which user can implement the transaction(s). But it is not mandatory that each and every private blockchain is permissioned	As anyone like user, developer or member is open to join the network of blockchain, regardless of his/her nationality or locality, which makes it impossible for authorities of network to shut down them. So, it can also be considered as highly censorship resistant in nature	In this case, for reference, the root hash and its API may be open to the public by using that API external users make a number of inquiries and obtain information related to blockchain	So, this type of blockchain is different from the other blockchains as it is not open to everyone however, it still provides features of blockchain that are, security, integrity, and transparency
Also private blockchains are more centralized in nature than the public one	Not more centralized as there is no decision taker of transactions in the network	It is semidecentralized in nature	After being decentralized, it makes possible to diminish the visibility of information on the network of blockchain
For example, supply chain management and monitoring and executing transactions, all are private.	For example, Litecoin and Ethereum blockchain systems and Bitcoin all are public.	Corda, Hyperledger, and Quorum are some examples of Consortium blockchain	Hybrid blockchain is captivating for regulated markets as it provides the advantages of both private and public blockchain together

6 The need for using a Blockchain in IoT

It must be highlighted that for every IoT scenario, the technology of blockchain is not always the finest solution. Sometimes, Directed Acyclic Graph or Traditional Data-bases Based Ledgers [16] may be considered as more suitable than blockchain for applications of IoT, particularly, to determine whether the usage of the technology of blockchain in Internet of Things is appropriate or not. So, for an IoT application, a developer should choose out the following features, which features are necessary:

- *Payment system*
  Some of the applications of Internet of Things may need to carry out commercial transactions with third parties, while several other applications of IoT may not. Furthermore, through the classical payment systems, the commercial transactions can still be performed, although it is essential to trust middleman or bank and usually pay fees of transaction.
- *Micro-transaction collection*
  To store the record of each transaction, to manage identifiability, for analyzing purposes or because of the technique of big data, which will be used later [17, 18], some of the IoT applications may require [19, 20]. A side chain may be proved useful in these kinds of circumstances. On the other side, there is no requirement to collect each recorded value in other applications of IoT. For instance, in the process of remote agricultural monitoring, which has expensive communications, it is very typical to make usage of nodes of IoT that get up in every 60 min to get the substantial info from sensors. In these types of conditions, all the data may be collected and stored by a local system and one time a day, the whole processed or refined information in the form of one transaction is transmitted by it.
- *Decentralization*
  In the absence of a trustworthy centralized system, applications of IoT demand decentralization. Still, many users trust blindly government banks or agencies and certain companies, so if there is a mutual trust then, there is no requirement of blockchain.
- *Public sequential transaction logging*
  There are some data that are collected by many IoT networks and require to be time stamped and stored in order. Nevertheless, these requirements may be easily achieved with the traditional data bases, specifically in those situations where attacks are infrequent and security is guaranteed.
- *P2P exchanges*
  In Internet of Things, most of the communications move from the nodes of IoT to the portals that transmit the data to a remote server or the cloud, except for particular applications, like in mist computing systems [21] or in intelligent swarms [22], interactions among peers at a level of node are not really very typical.
- *Robust distributed system*

  The requirement of this characteristic is not sufficient to explain the use of a blockchain, in the entity that maintains the disseminated system of computing, at least there also has to be an absence of trust (Fig. 6).

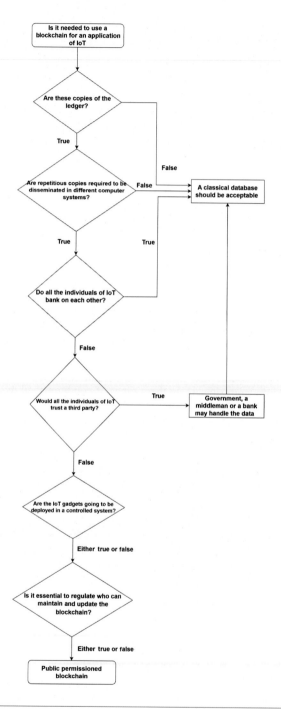

FIG. 6

Flow chart for determining the need, when to accommodate the blockchain technology to the particular requirements of an application of Internet of Things.

7 BIoT applications

Blockchain permits the IoT gadgets to bring the transparency and to boost-up security in IoT ecosystems. Also, blockchain provides a decentralized and scalable surrounding to the Internet of Things applications, platforms, and gadgets too.

As presented in Fig. 7, the blockchain applicability evolution is suggested by M. Swan. [2], which is started from bitcoin, which further then evolved to smart contracts and finally to coordinated and efficient utilizations or applications.

The technology of blockchain can be used in many fields and areas apart from smart contracts and cryptocurrencies with the involvement of IoT applications [23] like intelligent automotive systems [24], cyber law [25], timestamping services [26], supply chain management [27], and smart living applications [28]. Now, description of some of the applications of blockchain with the involvement of Internet of Things applications is as follow:

- In the case of IoT agricultural applications, blockchain can be used. Blockchain coordinated with IoT has the capability to reshape the production industry of food that is from farm to grocery and from grocery to home. To boost up the food supply chain to a greater extent, installation of IoT sensors in the farms can help which will send the all data directly to the blockchain. One of the examples to understand this better is in [29], it is shown that an identifiable system for tracing the supplies of Chinese agrifood which is typically based on a blockchain and the use of Radio Frequency Identification (RFID) with an aim to boost up the quality and safety of food.

- Nowadays, automotive industries are using sensors, that are IoT enabled to develop fully automated vehicles, keeping in mind that in current period, digitization is experienced as a competitive demand. Thus the industry of automation is considered as an interesting field for applying the blockchain IoT. In this field, many users are enabled to exchange important information quickly and easily by just connecting IoT enabled vehicles with the decentralized network.

- In literature, moreover, applications of BIoT are also applied in the field of healthcare. In [30], for reference, an application that make use of both blockchain technology and IoT sensors to check public convenience and data integrity of records of temperature in the supply chain of medicines is presented. This examination is considered as fault finding in the shipment of medicinal drugs, to assure both their surrounding circumstances (that are, their relative humidity and

FIG. 7

Evolution of blockchain applicability.

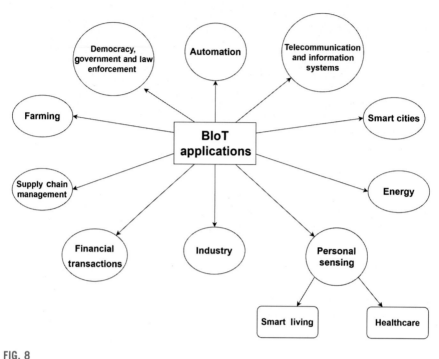

FIG. 8

Blockchain Internet of Things (BIoT) applications.

temperature) and quality. So, each and every transported parcel includes an IoT sensor that sends all the collected or recorded data to the blockchain where the received values will be checked by a smart contract to determine whether the values that are received are lying within the allowed range or not. Also the traceable and open nature of technology of blockchain can help to monitor the transportation of medicinal products or drugs from their origin to the supply chain's destination (Fig. 8).

8 Existing issues of applications of blockchain IoT

In current, many issues have to be faced by the emerging technologies in the environment of Internet of Things like RFID [31], 4G/5G broadband communications [32, 33], cyber physical systems (CPS) [34–36], or telemetry systems [37]. Addition of blockchain technology to the mixture signifies more technical as well operational necessities, and since the growth of blockchain IoT applications is considered as a complicated method that is affected by several conditions that are interconnected.

In the next coming sections, the main factors are detailed and represented in the Fig. 9.

FIG. 9

Components that complicate the growth of an application of blockchain IoT.

a. *Energy efficiency*

Energy efficiency is a key to allow a long-running deployment of node. But still, there are many blockchains that are defined by being power-desirous. In these types of situations, most of the utilization of power is due to the two main aspects:

- P2P communications:

Edge devices are required for P2P communications. These devices need to be powered on frequently, which could lead to wastage of energy [38, 39]. Energy efficient protocols for P2P networks [40] had proposed by M. Domb et al. [41], but still the issue has to be examined ahead for the particular situation of Internet of Things networks.

- Mining:

Due to the process of mining, Bitcoin like blockchains make usage of huge amount of electricity, which includes a consensus algorithm (Power of Work) [42].

b. *Privacy*

All the users of a blockchain are recognized by their public key or its hash, which means that privacy is not assured and that leads to sharing of all the transactions and it is possible for third parties to examine those transactions and deduce the real identities of the participants [43, 44]. In the ecosystem of IoT, privacy is even more complex than the thinking. In a blockchain the stored private user data can be revealed by the IoT gadgets, as privacy needs of blockchain vary from one country to another country [45].

Thus, in opposition to traditional online payments, that are commonly only see-able to a middleman (e.g., government and financial institutions) or to transacting parties, so the transactions which are transparent fostered by blockchain are an issue in terms of privacy.

Mixing techniques can also help to boost privacy. These kinds of techniques, from different IoT devices, collect transactions and other transactions or output events from distinct addresses that are not connected to the actual devices. There is no doubt that these techniques enhance privacy, but they cannot be considered as perfect, as they may be deanonymized through statistical disclosure attacks [46].

Through a zero-knowledge proving techniques like the ones used by Zerocoin [47], Zerocash [48], or Zcash [49], privacy can be raised. A method that permits for proving to a counterparty that a user known a certain information without revealing such that information [50] is known as a zero-knowledge proof. In the situation of applications of IoT, for authentication or during routine transactions, to prevent disclosing the identity of a device or a user, these proofs are not protected to irruptions (attacks).

Also, it must be remarked that the privacy-focused efforts are performed by several initiatives like Bytecoin [51] or Monero [52], which are based on CryptoNote [53]. CryptoNote is a protocol that makes use of ring signatures and whose transactions cannot be followed through the blockchain to determine who performed them.

The use of Homomorphic Encryption [54] is another feasible solution for maintaining privacy. This type of encryption, without revealing the unencrypted data to third-party IoT services allows those services to process a transaction. Variations of the Bitcoin protocol to make use of homomorphic commitments [55] have been suggested by M. Hamdi.

Finally, note that preceding discussed mechanisms need an appropriate number of computational resources. Therefore its applicability to resource-constrained IoT gadgets is recently limited.

c. *Blockchain infrastructure, size and bandwidth*

In respect of the infrastructure of blockchain, to make the blockchain work appropriately, certain elements are needed, including communication protocols, network administration or address management, decentralized storage, mining hardware. Part of these requirements is being satisfied by creating particular hardware for applications of blockchain and also by the industry progressively.

Techniques of blockchain compression should be more studied because as users store their transactions, correspondingly blockchains grow systematically, and in the coming period, most of the nodes of IoT would not be capable to manage even a small portion of a classical blockchain. Furthermore, it is noted that many nodes have to be record collected content in large extent that are of no use for them, which can be observed as a wastage of computational devices. By using light in weight nodes, which are capable to execute transactions, this issue can be resolved, but this approach needs the presence in the hierarchy of IoT of certain effective and dynamic nodes that would handle the blockchain for the resource constrained.

Another option is the usage of a mini blockchain [56]. This type of blockchain introduces the usage of an account tree, which collects and records the present state of every user of the blockchain, which leads to storing the only most current transaction on the blockchain with the account tree. Thus, in this case, size of the blockchain only increases on the addition of new users to the blockchain.

Moreover, according to the limitations of bandwidth of IoT networks, block size and transaction have to be scaled; for reference, a few large transactions may include huge payloads, while on the other side, many small transactions increment the utilization of energy that cannot be managed by some IoT devices.

d. *Security*

Generally, by an information system, three main necessities have to be fulfilled to guarantee its security:

- Availability: When required, data can be easily accessed.
- Confidentiality: The most delicate or private should be secured from illegal penetrations.
- Integrity: It assures that the info cannot be changed or modified or removed by any kind of illegal party. Also, it is included that, if in any case, an unauthorized party harms the data, it should be possible to undo the modifications.

The first feature of security that is, availability is the most forthright to be achieved by blockchains. However, availability can be accommodated through other types of irruptions. Majority attack that is also known as 51% attack is considered as the most feared attack. In this attack an individual miner can control the complete blockchain and execute transactions according to his/her wish [57].

Concerning confidentiality the part that is associated to the transaction data is linked with their secrecy, which has been previously examined in the preceding section. For a single user the key for managing confidentiality is an improved disposition of that user's private keys, since it is what an assailant requires in addition with the public key either to take something without permission like sensitive information from him/her or impersonate someone.

In respect of integrity, it must be implied that the foundation of a blockchain is created to record data that cannot be modified or changed once it is stored or collected. In the case of applications of IoT, the integrity of data is necessary, and it is commonly given by third parties. To prevent such a reliance, in [58] for applications of IoT based on cloud that make usage of the blockchain, a data integrity service framework is introduced, thus getting rid of the requirement of trusting such third-parties.

d. *Other appropriate issues*

- Usability
- Smart contract enforcement and autonomy
- Adoption rate
- Forks and versioning

9 Further issues and recommendations of blockchain in IoT

In spite of the interesting expected future and promising advantages of Blockchain Internet of Things, there are many serious issues in the both deployment and development of current and planned systems that will need more analysis:

- *Infrastructure of blockchain*
  It will be required to build an extensive trustworthy base or framework that can achieve all the necessities for the application of the technology of blockchain in Internet of Things networks. It is approached by several state of the art that introduce problems like faith, rely upon interdomain practices and monitoring. As a reference, to support use case of public interests, an infrastructure of blockchain should be set up by the governments.
- *Rapid field testing*
  In the upcoming period, for varied applications, various kinds of blockchains will require to be enhanced. Furthermore, in case, when the user wants to integrate a blockchain with IoT system(s), the first and important step is to check out which type of blockchain fits better with which IoT system [59]. So, there is a need to build an approach to test different blockchains and that mechanism should be divided into the two important stages:
  a. Standardization
  In this stage, after the detailed understanding of the markets, products, services, and supply chains, all the needs have to be examined and acknowledged. When a blockchain is established, it should be tested with the acknowledged procedure to validate that the established blockchain is working according to the requirement or not.
  b. Testing
  After the stage of standardization, in the situation of the testing stage, various procedures should be checked out in terms of efficiency of energy, security, latency, privacy, and blockchain's usability or capacity etc. [60].
- *Complex technical issues*
  With respect to security, stability needs, scalability, and cryptographic development of Blockchain Internet of Things (BIoT), still there are some issues

to be addressed. Furthermore, in the implementation of smart contracts, in capacity of transactions or in validation of protocols, design limitations are faced by the technology of blockchain. Additionally, to resolve the tendency of centralized approach [61], some methods should be introduced.

- *Governance, legal, organizational and regulatory aspects*

   Framing the regulating surrounding is one of the toughest issues to unravel the feasible value of Blockchain IoT in addition of technical challenges. For example, to draw attention of investor(s) driven by the expected benefits, it is possible that the performance of blockchain are faked by their respective developers.

10 Conclusion

It is impossible for us to refuse the fact that several technological innovations and advancements performed a primary role and made revolutionary transformations in our lives. Among different technologies, here comes a technology Internet of Things. With the approach of data naming in Internet of Things (IoT), a network solution was born, which is Named Data Networking of Things (NDNoT). Here, we illustrated various kinds of security irruptions occur in the network of NDNoT and how those irruptions reduce the functioning of Named Data Networking of Things. Furthermore, for Named Data Networking of Things, a solution of blockchain technology is introduced. As the Internet was significant 25 years ago, there is no doubt that the technology of blockchain is as significant now. And, it is proved that blockchain can provide IoT a platform for sharing trustworthy data that contravene nontrusted organizations. From the illustrations of the preceding sections, we can conclude that, still technology of Blockchain based Internet of Things is on its rising stage and, other than the earliest deployments and advancements of BIoT, there is a need of further technological research developments with the copartnership of governments and organizations to address the conspicuous requirements.

References

[1] IBM, Executive report: "Device democracy: Saving the future of the Internet of Things." (2015).

[2] M. Swan, Blockchain: Blueprint for a New Economy, first ed., O'Reilly Media, 2015.

[3] Gartner, Report: "Forecast: The Internet of Things, Worldwide, 2013" (2013).

[4] S. Landau, Highlights from making sense of Snowden, part II: what's significant in the NSA revelations, IEEE Secur. Priv. 12 (1) (2014) 62–64.

[5] T. Varshney, N. Sharma, I. Kaushik, B. Bhushan, Authentication & encryption based security services in blockchain technology. in: 2019 International Conference on Computing, Communication, and Intelligent Systems (ICCCIS), 2019. https://doi.org/10.1109/icccis48478.2019.8974500.

[6] M. Xie, I. Widjaja, H. Wang, Enhancing cache robustness for content-centric network-ing, in: INFOCOM, 2012 Proceedings IEEE, 2013, pp. 2426–2434.

[7] D. Kim, J. Bi, A.V. Vasilakos, I. Yeom, Security of cached content in ndn, IEEE Trans. Inf. Forensics Secur. 12 (12) (2017) 2933–2944.

[8] T. Mick, R. Tourani, S. Misra, Laser: lightweight authentication and secured routing for ndn iot in smart cities, IEEE Internet Things J. PP(99) (2017) 1.

[9] P. Afanasyev, I. Mahadevan, M.E. Uzun, L. Zhang, Interest flooding attack and coun-termeasures in named data networking, in: Ifip NETWORKING Conference, 2013, pp. 1–9.

[10] W. Shang, A. Bannis, T. Liang, Z. Wang, Y. Yu, A. Afanasyev, J. Thompson, J. Burke, B. Zhang, L. Zhang, Named data networking of things (invited paper), in: IEEE First International Conference on Internet-of-Things Design and Implementation, 2016, , pp. 117–128.

[11] K. Zhu, W. Zhi, X. Chen, L. Zhang, Socially motivated data caching in ultra-dense small cell networks, IEEE Netw. 31 (4) (2017) 42–48.

[12] Y.J. Kim, V. Kolesnikov, M. Thottan, Resilient end-to-end message protection for large-scale cyber-physical system communications, in: IEEE Third International Conference on Smart Grid Communications, 2012, , pp. 193–198.

[13] B. Vieira, E. Poll, A security protocol for information-centric networking in smart grids, in: ACM Workshop on Smart Energy Grid Security, 2013, , pp. 1–10.

[14] Ali, M., Uzmi, Z. A. "CSN: a network protocol for serving dynamic queries in large-scale wireless sensor networks", in Proceedings of the Second Annual Conference on Commu-nication Networks and Services Research, Fredericton, Canada, 21 May 2004.

[15] S. Landau, Making sense from Snowden: what's significant in the NSA surveillance rev-elations, IEEE Secur. Priv. 11 (4) (2013) 54–63.

[16] IOTA's official web page, Available online, https://www.iota.org (Accessed on 10 April 2018).

[17] H. Cai, B. Xu, L. Jiang, A.V. Vasilakos, IoT-based big data storage systems in cloud computing: perspectives and challenges, IEEE Internet Things J. 4 (1) (2017) 75–87.

[18] M. Marjani, F. Nasaruddin, A. Gani, A. Karim, I.A.T. Hashem, S. Siddiqa, I. Yaqoob, Big IoT data analytics: architecture, opportunities, and open research challenges, IEEE Access 5 (2017) 5247–5261.

[19] R. Tiwari, N. Sharma, I. Kaushik, A. Tiwari, B. Bhushan, Evolution of IoT & data analytics using deep learning. in: 2019 International Conference on Computing, Communication, and Intelligent Systems (ICCCIS), 2019. https://doi.org/10.1109/icccis48478.2019. 8974481.

[20] R.K. Lomotey, J. Pry, S. Sriramoju, E. Kaku, R. Deters, Wearable IoT data architecture, in: Proceedings of the IEEE World Congress on Services (SERVICES), Honolulu, United States, 25–30 June, 2017.

[21] J.S. Preden, K. Tammemäe, A. Jantsch, M. Leier, A. Riid, E. Calis, The benefits of self-awareness and attention in fog and mist computing, Computer 48 (7) (July 2015) 37–45.

[22] T. Gui, C. Ma, F. Wang, D.E. Wilkins, Survey on swarm intelligence based routing pro-tocols for wireless sensor networks: an extensive study, in: Proceedings of the IEEE International Conference on Industrial Technology (ICIT), Taipei, Taiwan, 14–17 March, 2016.

[23] A. Back, M. Corallo, L. Dashjr, M. Friedenbach, G. Maxwell, A. Miller, A. Poelstra, J. Timón, P. Wuille, Enabling Blockchain Innovations with Pegged Sidechains, Avail-able online: https://www.blockstream.com/sidechains.pdf (Accessed on 10 April 2018).

[24] M. Conoscenti, A. Vetrò, J.C. De Martin, Blockchain for the Internet of things: a systematic literature review, in: Proceedings of the IEEE/ACS 13th International Conference of Computer Systems and Applications (AICCSA), Agadir, Morocco, 29 November–2 December, 2016.

[25] Y. Zhang, J. Wen, An IoT electric business model based on the protocol of bitcoin, in: Proceedings of the 18th International Conference on Intelligence in Next Generation Networks, Paris, France, 17–19 February, 2015.

[26] B. Gipp, N. Meuschke, A. Gernandt, Decentralized trusted timestamping using the crypto currency bitcoin, in: Proceedings of the iConference, Newport Beach, United States, 24–27 March, 2015.

[27] D. Han, H. Kim, J. Jang, Blockchain based smart door lock system, in: Proceedings of the 2017 International Conference on Information and Communication Technology Convergence (ICTC), Jeju Island, South Korea, December 2017, , pp. 1165–1167.

[28] A. Wright, P. De Filippi, Decentralized Blockchain Technology and the Rise of Lex Cryptographia, Available online: https://ssrn.com/abstract=2580664, March 2015 (Accessed on 10 April 2018).

[29] F. Tian, An agri-food supply chain traceability system for China based on RFID & blockchain technology, in: Proceedings of the 13th International Conference on Service Systems and Service Management (ICSSSM), Kunming, China, 24–26 June, 2016.

[30] T. Bocek, B.B. Rodrigues, T. Strasser, B. Stiller, Blockchains everywhere—a use-case of blockchains in the pharma supply-chain, in: Proceedings of the IFIP/IEEE Symposium on Integrated Network and Service Management (IM), Lisbon, Portugal, 8–12 May, 2017.

[31] S.J. Barro-Torres, T.M. Fernández-Caramés, M. González-López, C.J. Escudero-Cascón, Maritime freight container management system using RFID, in: Proceedings of the Third International EURASIP Workshop on RFID Technology, La Manga del Mar Menor, Spain, 6–7 September, 2010.

[32] P. Fraga-Lamas, L. Castedo-Ribas, A. Morales-Méndez, J.M. Camas-Albar, Evolving military broadband wireless communication systems: WiMAX, LTE and WLAN, in: Proceedings of the International Conference on Military Communications and Information Systems (ICMCIS), Brussels, Belgium, 23–24 May, 2016, , pp. 1–8.

[33] P. Fraga-Lamas, J. Rodríguez-Piñeiro, J.A. García-Naya, L. Castedo, Unleashing the potential of LTE for next generation railway communications, in: Proceedings of the 8th International Workshop on Communication Technologies for Vehicles (Nets4Cars/Nets4Trains/Nets4Aircraft 2015), Sousse, Tunisia, 6–8 May 2015, Lecture Notes in Computer Science, 9066 Springer, Berlin/Heidelberg, Germany, 2015, , pp. 153–164.

[34] P. Fraga-Lamas, D. Noceda-Davila, T.M. Fernández-Caramés, M. DíazBouza, M. Vilar-Montesinos, Smart pipe system for a Shipyard 4.0, Sensors 16 (12) (December 2016) 1–43 no. 2186.

[35] P. Fraga-Lamas, T.M. Fernández-Caramés, D. Noceda-Davila, M. VilarMontesinos, RSS stabilization techniques for a real-time passive UHF RFID pipe monitoring system for smart shipyards, in: Proceedings of the 2017 IEEE International Conference on RFID (IEEE RFID 2017), Phoenix, AZ, USA, 9–11 May, 2017.

[36] P. Fraga-Lamas, T.M. Fernández-Caramés, D. Noceda-Davila, M. DíazBouza, M. Vilar-Montesinos, J.D. Pena-Agras, L. Castedo, Enabling automatic event detection for the pipe workshop of the Shipyard 4.0, in: Proceedings of the 2017 56th FITCE Congress, Madrid, Spain, 14–16 September, 2017, , pp. 20–27.

[37] D.L. Hernández-Rojas, T.M. Fernández-Caramés, P. Fraga-Lamas, C.J. Escudero, Design and practical evaluation of a family of lightweight protocols for heterogeneous

sensing through BLE beacons in IoT telemetry applications, Sensors 18 (1) (57) (December 2017) 1–33.

[38] Z. Zhou, M. Xie, T. Zhu, W. Xu, P. Yi, Z. Huang, Q. Zhang, S. Xiao, EEP2P: an energy-efficient and economy-efficient P2P network protocol, in: Proceedings of the International Green Computing Conference, Dallas, United States, 3–5 November, 2014.

[39] L. Sharifi, N. Rameshan, F. Freitag, L. Veiga, Energy efficiency dilemma: P2P-cloud vs. datacenter, in: Proceedings of the IEEE 6th International Conference on Cloud Computing Technology and Science, Singapore, Singapore, 15–18 December, 2014.

[40] P. Zhang, B.E. Helvik, Towards green P2P: analysis of energy consumption in P2P and approaches to control, in: Proceedings of the International Conference on High Performance Computing & Simulation (HPCS), Madrid, Spain, 2–6 July, 2012.

[41] C.C. Liao, S.M. Cheng, M. Domb, On designing energy efficient WiFi P2P connections for Internet of things, in: Proceedings of the IEEE 85th Vehicular Technology Conference (VTC Spring), Sydney, Australia, 4–7 June, 2017.

[42] S. Meiklejohn, M. Pomarole, G. Jordan, K. Levchenko, D. McCoy, G.M. Voelker, S. Savage, A fistful of bitcoins: characterizing payments among men with no names, Commun. ACM 59 (4) (April 2016) 8693.

[43] M. Möser, R. Böhme, D. Breuker, An inquiry into money laundering tools in the bitcoin ecosystem, in: Proceedings of the APWG eCrime Researchers Summit, San Francisco, United States, 17–18 September, 2013.

[44] T. Varshney, N. Sharma, I. Kaushik, B. Bhushan, Architectural model of security threats & their countermeasures in IoT. in: 2019 International Conference on Computing, Communication, and Intelligent Systems (ICCCIS), 2019. https://doi.org/10.1109/icccis48478.2019.8974544.

[45] G. Danezis, A. Serjantov, Statistical disclosure or intersection attacks on anonymity systems, in: Proceedings of the 6th International Workshop on Information Hiding, Toronto, Canada, 23–25 May, 2004, , pp. 293–308.

[46] Zerocoin official web page, Available online, http://zerocoin.org (Accessed on 10 April 2018).

[47] Zerocash official web page, Available online, http://zerocash-project.org (Accessed on 10 April 2018).

[48] Zcash official web page, Available online, https://z.cash (Accessed on 10 April 2018).

[49] M. Schukat, P. Flood, Zero-knowledge proofs in M2M communication, in: Proceedings of the 25th IET Irish Signals & Systems Conference and China-Ireland International Conference on Information and Communications Technologies, Limerick, Ireland, 26–27 June, 2014.

[50] K. Peng, Attack against a batch zero-knowledge proof system, IET Inf. Secur. 6 (1) (March 2012) 1–5.

[51] Bytecoin's official web page, Available online, https://bytecoin.org (Accessed on 10 April 2018).

[52] Monero's official web page, Available online, https://getmonero.org (Accessed on 10 April 2018).

[53] CryptoNote's official web page, Available online, https://cryptonote.org (Accessed on 10 April 2018).

[54] C. Moore, M. O'Neill, E. O'Sullivan, Y. Doröz, B. Sunar, Practical homomorphic encryption: a survey, in: Proceedings of the IEEE International Symposium on Circuits and Systems (ISCAS), Melbourne, Australia, 1–5 June, 2014.

[55] H. Hayouni, M. Hamdi, Secure data aggregation with homomorphic primitives in wireless sensor networks: a critical survey and open research issues, in: Proceedings of the

IEEE 13th International Conference on Networking, Sensing, and Control (ICNSC), Mexico City, Mexico, 28–30 April, 2016.

[56] B.F. França, Homomorphic Mini-blockchain Scheme, Available online: http://cryptonite.info/files/HMBC.pdf, April 2015 (Accessed on 10 April 2018).

[57] D. Lukianov, Compact Confidential Transactions for Bitcoin, Available online: http://voxelsoft.com/dev/cct.pdf, December 2015 (Accessed on 10 April 2018).

[58] K. Lei, M. Du, J. Huang, T. Jin, Group chain: towards a scalable public blockchain in fog computing of IoT services computing. IEEE Trans. Serv. Comput. 13 (2) (2020) 252–262, https://doi.org/10.1109/tsc.2019.2949801.

[59] B. Liu, X.L. Yu, S. Chen, X. Xu, L. Zhu, Blockchain based data integrity service framework for IoT data, in: Proceedings of the IEEE International Conference on Web Services, Honolulu, United States, 25–30 June, 2017.

[60] E. Birrell, F.B. Schneider, Federated identity management systems: a privacy-based characterization, IEEE Secur. Privacy 11 (5) (2013) 36–48.

[61] M. Bedford Taylor, The evolution of bitcoin hardware, Computer 50 (9) (22 September 2017) 58–66.

A novel privacy-preserving healthcare information sharing platform using blockchain

11

Mohammad Jaber, Amin Fakhereldine, Mahdi Dhaini, and Ramzi A. Haraty

Department of Computer Science and Mathematics, Lebanese American University,
Beirut, Lebanon

Chapter outline

1 Introduction

We live in the era of information. Data have been increasingly computerized from paper into electronic form. Medical data are among the popular types of data that are being stored in an electronic form in medical institutions. Sharing medical data online (laboratory tests, x-ray images, health surveys, etc.) has various advantages for medical research and treatment. Effective sharing of medical records can be very beneficial for all parties involved. In fact, medical records are being fragmented, rather than being kept cohesive, due to the lack of coordinated data management and exchange [1]. Therefore the need of interoperability between patients and medical institutions is clear. In addition, a patient's data can be aggregated at one

Security and privacy issues in IoT devices and sensor networks. https://doi.org/10.1016/B978-0-12-821255-4.00011-0

repository instead of being scattered across several systems, none of which has a complete picture about the patient's case.

However, more barriers are posed due to the interoperability challenges between hospital systems and providers [2]. In 2009 HITECH, part of the American Recovery and Reinvestment Act, spent about 30 billion dollars to incentivize electronic health record (EHR) adoption by US healthcare providers [3], causing an increase in the use of EHRs from 9% in 2008 to 96% by 2015. However, shifting to EHRs imposed different challenges mainly related to security (data privacy, authentication, scalability, etc.). Consequently, research gates were opened to study secure storage and sharing of health data [4]. Motivated by the aforementioned issues, the need for secure medical data sharing systems is essential. In such systems the patient's interests should be highly considered. One of the patients' main concerns is the ability to get their medical records from different hospital databases, to avoid the high costs of being obliged to undertake the same tests and examinations a number of times due to the inability of querying such records stored in another hospital's private database.

Many research studies have been done to reach this target [2,5], and blockchain was found to be a very suitable solution for this issue. The immutability, verifiability, and decentralized nature are some attractive features that make blockchain an appropriate solution. In this framework, blockchain was considered to have an important role in improving interoperability in health data systems because of its characteristics of sharing and distribution [3,6,7]. It had a great impact on improving cryptocurrency systems after its successful application in Bitcoin. Research was recently directed toward deploying blockchain in several systems such as education, cybersecurity, and IoT [8–10]. Moreover, it can provide health data systems with several features such as data immutability and access rules. However, such an application faces different challenges like privacy, security, scalability, and access rules.

The potential success of solutions based on blockchain has been shown to be evident by the plans of financing blockchain applications by healthcare organizations [11]. Based on the statistics found in [11], it can be clearly seen how blockchain has been highly considered as a potential solution in the healthcare sector. In this paper, we propose a patient-driven medical data sharing system using blockchain technology. The main contributions in our proposed model can be described as follows: patients are considered the main participants where they can upload and retrieve their medical data to a secure database with the ability of granting access to healthcare providers to analyze these information. In this system the patient monitors all the actions performed on his/her data such as uploading, viewing, and sharing information and gets notified when malicious behavior is done on his/her behalf. This system tackles the security barriers imposed because of the interoperability between healthcare providers and the patients. The security issues are studied carefully whereby we provide integrity and privacy of data, authentication of users, scalability, and access control. We also deploy the blockchain technology in a way that meets the requirements of securing medical data rather than the openness that was applied in cryptocurrency such as Bitcoin.

The rest of the paper is devised as follows: in Section 2, we review prior art work. In Section 3, we present a brief overview of the blockchain technology. Section 4 describes the proposed system model. In Section 5, we analyze the work of our system and compare it with an existing model in the literature. Finally, Section 6 concludes our paper and introduces potential future work.

2 Literature review

The topics of healthcare systems and secure sharing of medical records have been studied extensively in the literature. Some previous works proposed novel algorithms to secure healthcare systems [12–17]. These algorithms were proposed for pattern discovery and to ensure the consistency and stabilization of medical system's databases to the correct state while. Gordon and Catalini [3] discuss two types of interoperability in healthcare systems: institution driven and patient driven. Al Omar et al. [18] present a privacy preserving platform for data in healthcare data management systems that became recently a target for cyberattackers. Linn and Koo [19] address the need of interoperability in the health sector between data providers, patients, and organizations and stress the importance of utilizing the large advancements in the IT sector that provide patients access to their health data and facilitate their engagement in their own health care. The attractive features of cloud computing such as scalability, pricing-flexibility, and availability make it a very good candidate for healthcare data storage [20–23]. However, due to the risks of cloud sourcing and the sensitive nature of healthcare data and medical records, additional security requirements should be considered when using cloud storage for healthcare data sharing and exchange [24].

In a medical context, Kish and Topol [25] proposes a blockchain based solution to solve the hypothetical key management issue. Previous works—that involved using blockchain for medical records—include [26,23]. Azaria et al. [2] are the first to introduce a fully functional blockchain-based prototype for sharing medical records. A blockchain application for medical information sharing is introduced in [27]. In the paper the authors propose a new business process and a blockchain-based medical information sharing platform. Results were encouraging as the proposed platform had many advantages compared with the other solutions. Most notably, it allows less pressure on the main chain than MedRec [2], because of the introduction of a middleware layer [27]. Xia et al. [28] propose MeDShare, a system that addresses the issue of medical data sharing among medical big data keepers in a trustless environment. The proposed system is a blockchain-based solution for sharing medical data among cloud service providers while providing data access control, provenance, and auditing. Fan et al. [5] propose MedBlock, a blockchain-based system, to share electronic medical records among authorized users. Zhang et al. [29] apply blockchain technology to propose an FHIRChain architecture for clinical data sharing between different providers.

Moreover, to face limitations in healthcare sharing systems, Tanwar et al. [30] discusses several solutions for the current problem using blockchain technology. The proposed system eliminates the single authority and the single point of failure. To achieve better results, performance metrics such as latency, throughput, and round trip time have been optimized and enhanced. Trying to achieve both privacy and security for patient's data, Brunese et al. [31] proposes a new method aimed to protect exchanged information among hospitals based on blockchain technology. This method exploits formal verification techniques using equivalence checking to authenticate each host and to ensure that the received information is not changed or altered by an attacker. A patient centric healthcare data management system using blockchain technology to attain privacy is presented in Omar et al. [32]. Regarding security constraints the authors use cryptographic functions to encrypt patient's data. In Tripathi et al. [33] a new blockchain-based smart healthcare system framework is presented to overcome the barriers related to secure and private sharing of healthcare data.

In this paper, we present a medical data sharing system model based on block-chain technology. Our model meets with some previous work [2,5,27,29] in the sense of securing the sharing of medical information and the tamper proof nature. Our model though is different than some previous work in the literature as it considers the patients' interests first and foremost. Our proposed model is patient centric—all the exchange and sharing of medical records take place only after the patient's approval. The patient decides who can view and who can share his data.

3 Overview of blockchain

Blockchain was first introduced in Satoshi [34] as a framework for Bitcoin. Block-chain can be defined as a distributed ledger technology that allows the storing and sharing of data in a decentralized and immutable manner through a network of distributed peer-to-peer members. In this technology, blocks of data are connected through chains, thus forming a blockchain [33]. The blockchain structure is shown in Fig. 1 where each data block has a unique identifier, that is, a hash and also

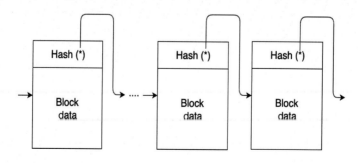

FIG. 1

Blockchain structure.

contains the hash of the previous data block. The main components and services that Blockhain comprises are a shared ledger, distributed peer-to-peer network, consensus mechanisms, and miners. The distributed ledger can be considered as data structure where all transactions in the network are recorded. Instead of sharing records in a conventional centralized manner through a consolidated authority, the shared ledger is maintained in a decentralized manner such that every transaction is processed by every node or participant on the distributed network.

The transactions are broadcast through the distributed peer-to-peer network to different network members to update data. However, the distributed ledger is only updated after consensus is reached; that is, after an agreement is reached between different nodes on the order and state of transactions. Consensus is maintained through some validation rules called consensus rules. These rules are provided by consensus algorithms. The proof-of-work (PoW) algorithm used in the Bitcoin blockchain is one of the most well-known consensus algorithms. PoW ensures verifiability and authentication by requiring a complicated computational puzzle to be solved [35]. However, this mechanism requires long processing time and high energy consumption when compared with other algorithms such as proof of stake (PoS). The PoS protocols, used in Peercoin [36], were introduced as an energy efficient alternative approach to PoW. Different consensus approaches have been based on the blockchain type. In 2016 Intel proposed the concept of proof of elapsed time (PoET) that has been part of the Hyperledger Sawtooth blockchain framework [37]. The Practical Byzantine Fault Tolerance (PBFT) was utilized by Hyperledger Fabric [38]. The Delegated Proof of Stake (DPOS), a representative democratic approach, was utilized by Bitshares [39]. The Ripple protocol is implemented by Ripple [40]. The main differences between the different consensus approaches include managing node identity, energy efficiency, and tolerated power of adversary.

Miners, also known as network maintainer nodes or nodes of the networks, link the blocks together to form the blockchain. They also compete in solving the complicated computational process required by the consensus algorithm. These nodes group the available transactions into blocks after validating them [29]. The history of transactions is secured through cryptography where both symmetric and asymmetric cryptographic algorithms have been used to secure the transactions history [2,5]. Another important component of blockchain is smart contracts. A smart contract can be defined as an agreement between parties where the agreed terms are self-enforced, although the parties do not trust each other [35]. In the context of blockchain, smart contracts allow the operations of transactions without depending on any other third party or trusted authority. They allow code to execute independently after some conditions are met (i.e., after the agreement between the involved parties). Smart contracts can also store information and update their state where their operations are stored as transactions, and are thus verified by the miners (e.g., in Ethereum blockchain) [29].

Blockchain got its decentralized nature and immutability from its features and components. The design and characteristics of its architecture nourished blockchain with security, audibility, robustness, and transparency [37,41]. These attractive properties have encouraged the deployment of blockchain technology in different sectors.

Aside from cryptocurrencies, blockchain has been explored in different applications such as healthcare systems [2,25,27,29], integrity verification [42,43], Internet of Things or IoT [44–46], energy sector [47,48], and data management [49,50]. In the next section, our proposed blockchain based healthcare model is presented.

4 System model

In this section, we present an overview of our proposed model where we present the main participants in our system and the main structural components. After that, we define the operations that can be performed by each of the participants with detailed explanation for the functional and security steps.

4.1 Overview

Fig. 2 shows our models' architecture, which is considered a solution to a wide range of healthcare sharing systems and is fully controlled by the patient. For simplicity, we consider that all the medical data provided to any patient is signed by authorized parties such as a hospital, laboratory, or a medical doctor.

As shown in Fig. 2, our model consists of four main participants:

1. Patient: has the main role in our system. The patient has full access and control over all his/her medical information. Patients can upload their medical information to/from the blockchain and grant and revoke access to providers. We can differentiate two types of accesses in our model: view access, which is granted by patients to authorized providers to view their medical information, and share access, which is granted by patients to providers to share their medical information with other providers. When the data are uploaded successfully, patients get notified and receive the address of their medical information in the database. Moreover a notification is received by the patient whenever a provider viewed (or attempted to view) his/her medical information alongside with the data viewed.

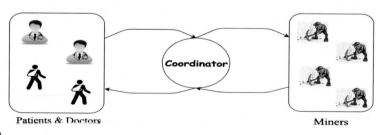

Patients & Doctors   Miners

FIG. 2

Architectural components of our model.

2. Provider: can be a doctor, a hospital, a medical laboratory, or an insurance company. A provider is granted access by his/her patient to access his/her medical information, add medical data to the blockchain on behalf of the patient, or even share the medical data with an authorized provider.
3. Miner: is a network component that has a copy of the blockchain database and can connect with the smart contract to verify and validate transactions (adding, viewing, or sharing medical data) based on the network consensus.
4. Coordinator: is the middle layer in our model. Whenever a provider or a patient needs to access or write data to the blockchain, the request is sent to the coordinator. The coordinator broadcasts the request to the miners to verify the transaction and receive their response. Based on the consensus algorithm, the coordinator decides to accept or reject the request based also on a new consensus algorithm. The sender (patient or provider) will receive a notification with all the details required (rejection message, address of saved data, or identities of providers who accessed the data).

In addition, three main components are needed in our model:

1. Secure database: all medical data related to a patient such as lab tests, MRIs, x-ray images, and documents are stored encrypted in the secure database at a certain address to ensure privacy and authenticity of the information. The encryption of these addresses will be stored in the blockchain, which proved better than saving all the medical information to save more space. This helps in both security and scalability where only encrypted addresses of health data are stored in the blockchain rather that storing the whole data. Security wise, if an attacker gained access to the blockchain, she/he will be able to view useless encrypted address instead of viewing the whole clear information.
2. Blockchain: a decentralized database based on the Ethereum platform, used to save public keys representing user identities (provider and patient) for identity and tamper proofing. Moreover, it contains encrypted pointers to the medical data in the secure database. The decentralized nature of the blockchain combined with digitally signed transactions ensure that an adversary cannot pose as the user, or corrupt the network for that would imply the adversary forged a digital signature or gained control over most of the networks' resources, which seems impossible. Similarly an adversary would not be able to learn anything from the shared public ledger as only encrypted pointers would be contained within the transactions.
3. Smart contract: is an algorithm or an application that is used to identify users (patients or providers). It has full access on the directory of public keys and their corresponding private keys. Also, it has full knowledge about data access control provided by each patient to his/her corresponding providers. In other words, it identifies for each patient which providers are allowed to view or share his/her data.

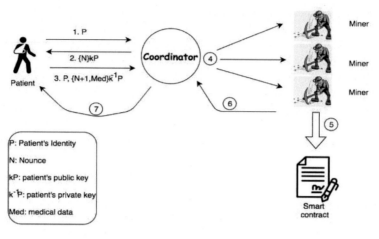

FIG. 3

Patient uploading data.

4.2 Patient uploading medical data

The procedure of uploading medical information to the blockchain by a patient is summarized in Fig. 3.

The steps proceed as follows:

1. The patient sends his/her identity to the coordinator to open a connection.
2. The coordinator replies with a nonce N encrypted by the patient's public key so that only the corresponding patient can decrypt it.
3. The patient replies with his/her identity sent in the clear alongside with a new nonce N + 1 and the medical data she/he wants to upload all signed by his/her private key to verify authenticity that the data are sent by him/her.
4. The coordinator receives the transaction from the user and broadcasts it to the miners.
5. The miners collaborate with the Smart contract to check if the provided data are signed with the private key, which corresponds to the public key provided earlier in the transaction.
6. Miners will reply to the coordinator with the corresponding response.
7. The coordinator will then take the decision whether to accept the transaction or reject it based on the consensus algorithm and the responses received from the miners. If the transaction was accepted, the medical data will be stored in the secure database (encrypted by the patient's public key) at a certain address. This address will be saved in the blockchain (encrypted by the patient's public key), and the patient will receive a notification stating that the procedure was successfully completed along with the address at which the data were saved (encrypted with the patient's public key). If the transaction was rejected, it means

that an attacker is attempting to upload medical information on behalf of the patient, so the patient will receive a message notifying him/her that an attempt to save medical data was rejected alongside with the public key of the attacker.

4.3 Provider uploading medical data of his/her patient

The procedure of uploading patient's medical information to the blockchain by a provider is summarized in Fig. 4. The steps proceed as follows:

1. The patient sends his/her identity to the coordinator to open a connection.
2. The coordinator replies with a nonce N encrypted by the providers public key so that only the corresponding provider can decrypt it.
3. The provider replies to the coordinator with his/her identity sent in the clear, alongside with a new nonce N + 1, the medical data he/she wants to upload, and the public key of the patient, all signed (encrypted) by his/her private key to verify that these data are sent by him/her.
4. The coordinator receives the transaction from the user and broadcasts it to the miners.
5. The miners collaborate with the Smart contract to check if (i) the provided data are signed with the private key corresponding to the public key of the provider and (ii) the provider is granted access to upload medical data on behalf of the patient.
6. Miners will reply to the coordinator with the corresponding response.

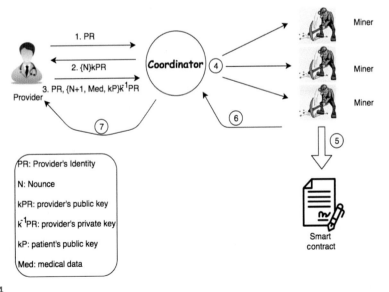

FIG. 4

Provider uploading data.

7. The coordinator takes the decision of whether to accept or reject the transaction according to the responses received from the miners and by applying the consensus algorithm. If the transaction is accepted, the medical data will be stored in the secure database (encrypted by the patient's public key) at a certain address. This address will be saved in the blockchain (encrypted by the patient's public key). The provider will receive a notification that the transaction was completed successfully alongside with the medical data address encrypted first by the patient's public key, and then encrypted with the providers' public key. The corresponding patient will also receive a message that data were saved to his/her blockchain specifying the provider identity. If the transaction is rejected, both the patient and the provider with the given public keys will be sent a notification message that an attempt to save medical data was rejected alongside with the reason (e.g., if the provider attempting to add medical data is not authorized by the corresponding patient).

4.4 Provider sharing medical data with another provider

The procedure of sharing medical information by two providers is summarized in Fig. 5. The steps proceed as follows:

1. The provider sends his/her identity to the coordinator to open a connection.
2. The coordinator replies with a nonce N encrypted by the providers public key so that only the corresponding provider can decrypt it.

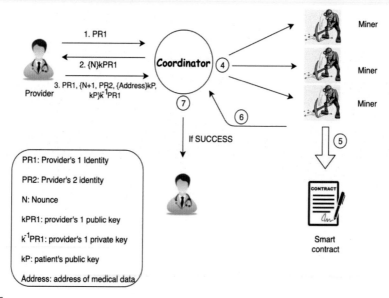

FIG. 5

Provider sharing data with another provider.

3. The provider replies to the coordinator with his/her identity sent in the clear, alongside with a new nonce N + 1, the identity of Provider2, and the encrypted data address (by patient's public key) all signed (encrypted) by his/her private key to verify that these data are sent by him/her.
4. The coordinator receives the transaction from the provider and broadcasts it to the miners.
5. The miners collaborate with the Smart contract to check if (i) the provided data are signed with the private key corresponding to the public key of the provider and (ii) the provider is granted access to share the medical data of the provided patient.
6. Miners will reply to the coordinator with the corresponding response.
7. The coordinator takes the decision either to accept or reject the transaction according to the responses received from the miners and by applying the consensus algorithm. If the transaction is accepted, then Provider2 will receive the address of the data (encrypted by the patients' public key). Provider1 will receive a message notifying that the medical data are sent successfully to Provider2, and the patient will receive a transaction log. If the transaction is rejected, the patient and Provider1 with the given public keys will be receiving a message notifying them that an attempt to share medical data with Provider2 was rejected with the reason.

4.5 Provider querying medical data of patient

The procedure of querying patient's medical information by provider is summarized in Fig. 6. The steps proceed as follows:

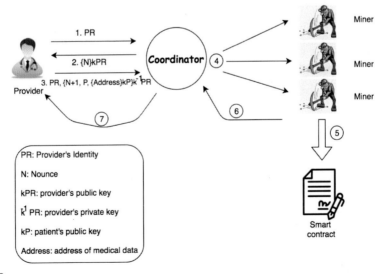

FIG. 6

Provider querying medical data.

1. The provider sends his/her identity to the coordinator to open a connection.
2. The coordinator replies with a nonce N encrypted by the providers public key so that only the corresponding provider can decrypt it.
3. The provider replies with his/her identity sent in the clear, alongside with a new nonce N + 1, the identity of the patient, and the encrypted data address all signed by the providers' private key, to the coordinator.
4. The coordinator receives the transaction from the provider and broadcasts it to the miners.
5. The miners collaborate with the Smart contract to check if (i) the provided data are signed with the private key corresponding to the public key of the provider and (ii) that the provider has access to view data of the provided patient.
6. Miners will reply to the coordinator with the corresponding response.
7. The coordinator takes the decision either to accept or reject the transaction according to the responses received from the miners and by applying the consensus algorithm. If the transaction is accepted, the provider will receive the needed data encrypted by his/her public key, and the patient will get a message notifying him/her that the specified provider has successfully viewed his/her data. If the transaction is rejected, both the patient and the provider with the given public keys will be receiving a message notifying him/her that an attempt to view data was rejected with the reason.

4.6 Patient updating access given to provider

The procedure of updating access granted to a provider is summarized in Fig. 7. The steps proceed as follows:

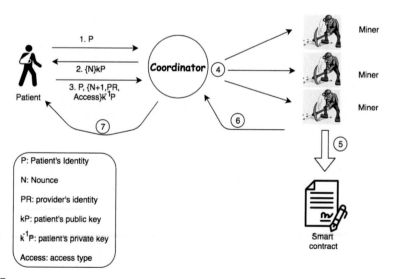

FIG. 7

Patient updating access given to provider.

1. The patient sends his/her identity to the coordinator to open a connection.
2. The coordinator replies with a nonce N encrypted by the patient's public key so that only the corresponding patient can decrypt it.
3. The patient replies with his/her identity sent in the clear, alongside with a new nonce N + 1, the identity of the provider, and the access type all signed by the patient's private key, to the coordinator.
4. The coordinator receives the transaction from the provider and broadcasts it to the miners.
5. The miners collaborate with the Smart contract to check if (i) the provided data are signed with the private key corresponding to the public key of the patient and (ii) that the access type is predefined.
6. Miners will reply to the coordinator with the corresponding response.
7. The coordinator takes the decision of whether to accept or reject the transaction according to the responses received from the miners and by applying the consensus algorithm. If the transaction is accepted, the provider will be informed that the transaction was successful and the corresponding provider will be notified of the new access granted to him/her. If the transaction is rejected, the patient will be sent a message notifying him/her that an attempt to update access was rejected with the reason.

5 Analysis

As discussed in previous sections, we introduced a patient-driven sharing system for healthcare information where patients can perform and monitor all actions related to their medical data such as uploading/retrieving data by them or their healthcare providers and providing access control. In this section, we analyze our model and compare it with an existing model, which is FHIRChain [29]. The reason behind picking this model for comparison is that it is a blockchain-based model designed to meet the requirements for healthcare sharing systems defined by the Office of the National Coordinator for Health Information Technology (ONC). ONC serves in overseeing the development of secure healthcare systems as a division of the Office of the Secretary within the United States Department of Health and Human Services [29].

In our model, as in FHIRChain, we take into consideration several barriers that could limit healthcare systems such as medical identity theft that could affect security and privacy of data, lack of trust between healthcare providers, and scalability that could affect the system response time. The following three requirements summarize the main requirements needed for having a secure, interoperable, and privacy-preserving healthcare system based on the ONC recommendations: authenticating participants, secure storage and exchange of data, and ensuring permissioned access to the stored data.

To provide authentication of participants, we used public key cryptography in our model where each participant has a pair of public and private keys. The public key is considered a public identity as it is known to every other participant. The private key

is computationally hard to obtain from the public key and is used for signature purposes to provide privacy and confidentiality of information especially while data
exchange. The same mechanism was used in FHIRChain for authentication purposes.

Moreover, to maintain secure storage and exchange of data, the medical data in
our model are stored in a secure database away from the blockchain. We store on the
blockchain the addresses at which the data are stored in the secure database,
encrypted by the public key of the patient. A similar approach is used in FHIRChain.
It stores the medical data in a secure database and keeps encrypted addresses in the
blockchain. However, in our model, we keep the medical data encrypted by the
patient's public key to provide more security. By this approach, we are supporting
scalability as well since storing the encrypted addresses on the blockchain is much
lighter than storing the actual data. Keeping in mind the decentralized nature of
blockchain, it is not feasible to store the entire medical information on it, and creating
several copies of it.

Concerning permissioned data access, the data are mainly signed with the private
key of the sender and encrypted with the public key of the receiver. This ensures that
the sender and receiver dealing with these data are really the intended ones. However, to keep the patient in the middle of the process, the smart contract makes sure
that any provider who would be accessing the data is actually given authority by the
patient to access these data. In FHIRChain, they employed a similar approach of
signing then encrypting, but without taking into consideration the access control
rules provided by the patient.

Based on this comparison and the steps explained in Section 4 to perform any
action in the system, we can realize how our model provided a secure patient-driven
healthcare platform. The patient gets notified whenever any provider accesses his/
her data or manipulates his/her access control rules. The patient also gets notified
on the spot whenever any user tries to perform any malicious behavior on his/her
behalf. This can actually help in achieving trust between the system participants.
Data security is achieved as well where the secure database holds encrypted medical
information and the blockchain holds encrypted data addresses. This means that,
even if an attacker was able to gain access to them (which is technically hard),
he/she would get encrypted data instead of clear ones. The fact that these data are
encrypted by the patient's public key provides integrity to the system.

6 Conclusion

The utilization and use of blockchain technology in different domains such as education, healthcare, and finance have been rapidly increasing. The need for secure
medical sharing systems is essential in this era of information. In this paper, we
explored the unique advantages and features of blockchain to propose a novel healthcare information sharing platform based on blockchain. Our patient-driven proposed
model is distinguished from previous work in literature such as FHIRChain [29]
where the patient is the main participant in our model. Analysis proved that our

model provides authenticity and confidentiality and is tamper proof. In our model, we improved the integrity, scalability, and storage when compared with FHIRChain. As for future work, we will look to enhance our system by introducing body sensors (IoT sensors) and migrating them in our model to receive health information of the patient directly. These data will be stored securely in the blockchain to be shared upon the approval of the patient. Further investigations are needed to check the potential success of this suggested work.

References

[1] K.D. Mandl, P. Szolovits, I.S. Kohane, Public standards and patients' control: how to keep electronic medical records accessible but private, BMJ 322 (7281) (2001) 283–287.

[2] A. Azaria, A. Ekblaw, T. Vieira, A. Lippman, MedRec: using blockchain for medical data access and permission management, Int. Conf. Open Big Data (2016) 25–30.

[3] W.J. Gordon, C. Catalini, Blockchain technology for healthcare: facilitating the transition to patient-driven interoperability, Comput. Struct. Biotechnol. J. 16 (2018) 224–230.

[4] S.R. Islam, D. Kwak, M.H. Kabir, M. Hossain, K.-S. Kwak, The internet of things for health care: a comprehensive survey, IEEE Access 3 (2015) 678–708.

[5] K. Fan, S. Wang, Y. Ren, H. Li, Y. Yang, MedBlock: efficient and secure medical data sharing via blockchain, J. Med. Syst. 42 (8) (2018) 136, https://doi.org/10.1007/s10916-018-0993-7.

[6] M. Ali, J. Nelson, R. Shea, M. Freedman, Blockstack: Design and Implementation of a Global Naming System with Blockchains, 25, 2019, p. 2.

[7] S. Jiang, J. Cao, H. Wu, Y. Yang, M. Ma, J. He, BlocHIE: a BLOCkchain-based platform for Healthcare Information Exchange, in: 2018 IEEE International Conference on Smart Computing (SMARTCOMP), Taormina, 2018, pp. 49–56.

[8] M. Turkanovi, M. Hlbl, K. Koi, M. Heriko, A. Kamiali, EduCTX: a blockchain based higher education credit platform, IEEE Access 6 (2018) 5112–5127.

[9] J. Chen, X. Ma, M. Du, Z. Wang, A blockchain application for medical information sharing, in: TEMS-ISIE 2018–1st Annual International Symposium on Innovation and Entrepreneurship of the IEEE Technology and Engineering Management Society, 2018, pp. 1–7.

[10] S. Huh, S. Cho, S. Kim, Managing IoT devices using blockchain platform, in: 2017 19th International Conference on Advanced Communication Technology (ICACT), Bongpyeong, 2017, pp. 464–467.

[11] The Economist Intelligence Unit of IBM Institute for Business Value, Healthcare rallies for blockchains: keeping patients at the center. Healthcare and blockchain executive report, Available online, http://www.ibm.biz/blockchainhealth, 2017.

[12] O. El Zarif, R.A. Haraty, Toward information preservation in healthcare systems, in: Next Gen Tech Driven Personalized Med & Smart Healthcare, Innovation in Health Informatics, Academic Press, ISBN: 9780128190432, 2020, pp. 163–185, https://doi.org/10.1016/B978-0-12-819043-2.00007-1.

[13] R.A. Haraty, S. Kaddoura, A.S. Zekri, Recovery of business intelligence systems: towards guaranteed continuity of patient centric healthcare systems through a matrix-based recovery approach, Telematics Inform. 35 (4) (2018) 801–814, https://doi.org/10.1016/j.tele.2017.12.010.

[14] R.A. Haraty, M. Zbib, M. Masud, Data damage assessment and recovery algorithm from malicious attacks in healthcare data sharing systems, J. Peer-to-Peer Netw. Appl. 9 (5) (2016) 812–823, https://doi.org/10.1007/s12083-015-0361-z.

[15] R.A. Haraty, Innovative mobile e-healthcare systems: a new rule-based cache replacement strategy using least profit values, Mob. Inf. Syst. 2016 (2016), https://doi.org/10.1155/2016/6141828.

[16] R.A. Haraty, M. Dimishkieh, M. Masud, An enhanced k-means clustering algorithm for pattern discovery in healthcare data, Int. J. Distrib. Sens. Netw. 11 (6) (2015), https://doi.org/10.1155/2015/615740.

[17] S. Kaddoura, R.A. Haraty, A. Zekri, M. Masud, Tracking and repairing damaged healthcare databases using the matrix, Int. J. Distrib. Sens. Netw. 11 (11) (2015), https://doi.org/10.1155/2015/914305.

[18] A. Al Omar, M.S. Rahman, A. Basu, S. Kiyomoto, A blockchain based privacy preserving platform for healthcare data, in: International Conference on Security, Privacy and Anonymity in Computation, Communication and Storage, Springer, 2018, pp. 534–543.

[19] L.A. Linn, M.B. Koo, Blockchain for health data and its potential use in health IT and health care related research, in: ONC/NIST Use of Blockchain for Healthcare and Research Workshop. Gaithersburg, Maryland, United States: ONC/NIST, 2016.

[20] Y.Y. Chen, J.C. Lu, J.K. Jan, A secure EHR system based on hybrid clouds, J. Med. Syst. 36 (5) (2012) 3375–3384.

[21] Y. Yang, X. Li, N. Qamar, P. Liu, W. Ke, B. Shen, Z. Liu, MedShare: a novel hybrid cloud for medical resource sharing among autonomous healthcare providers, IEEE Access 6 (2018) 46949–46961.

[22] Y. Yang, Z. Quan, P. Liu, D. Ouyang, X. Li, MicroShare: privacy-preserved medical resource sharing through MicroService architecture, Int. J. Biol. Sci. 14 (2018) 907–919.

[23] B. Shen, J. Guo, Y. Yang, MedChain: efficient healthcare data sharing via blockchain, Appl. Sci. 9 (2019) 1207, https://doi.org/10.3390/app9061207.

[24] H. Abrar, S.J. Hussain, J. Chaudhry, K. Saleem, M.A. Orgun, J. Al-Muhtadi, C. Valli, Risk analysis of cloud sourcing in healthcare and public health industry, IEEE Access 6 (2018) 19140–19150.

[25] L.J. Kish, E.J. Topol, Unpatients – why patients should own their medical data, Nat. Biotechnol. 33 (9) (2015) 921–924.

[26] Medical records project wins top prize at blockchain hackathon, [online]. November 15. Available:http://www.coindesk.com, 2015.

[27] J.R. Douceur, The Sybil attack, in: P. Druschel, F. Kaashoek, A. Rowstron (Eds.), Peer-to-Peer Systems. IPTPS 2002, Lecture Notes in Computer Science, Vol. 2429, Springer, Berlin, Heidelberg, 2002.

[28] Q. Xia, E.B. Sifah, K.O. Asamoah, J. Gao, X. Du, M. Guizani, MeDShare: trust-less medical data sharing among cloud service providers via blockchain, IEEE Access 5 (2017) 14757–14767.

[29] P. Zhang, J. White, D.C. Schmidt, G. Lenz, S.T. Rosenbloom, FHIRChain: applying blockchain to securely and scalably share clinical data, Comput. Struct. Biotechnol. J. 16 (2018) 267–278, https://doi.org/10.1016/j.csbj.2018.07.004.

[30] S. Tanwar, K. Parekh, R. Evans, Blockchain-based electronic healthcare record system for healthcare 4.0 applications, J. Inf. Secur. Appl. 50 (2020) 102407, https://doi.org/10.1016/j.jisa.2019.102407.

[31] L. Brunese, F. Mercaldo, A. Reginelli, A. Santone, A blockchain based proposal for protecting healthcare systems through formal methods, Procedia Comput. Sci. 159 (2019) 1787–1794.

[32] A. Omar, M.Z.A. Bhuiyan, A. Basu, S. Kiyomoto, M.S. Rahman, Privacy-friendly platform for healthcare data in cloud based on blockchain environment, Futur. Gener. Comput. Syst. 95 (2019) 511–521.

[33] G. Tripathi, M.A. Ahad, S. Paiva, S2HS – a blockchain based approach for smart healthcare system, Healthcare 8 (1) (2019) 100391.

[34] N. Satoshi, Bitcoin: a peer-to-peer electronic cash system, Consulted 1 (2008) 1–9.

[35] F. Casino, T.K. Dasaklis, C. Patsakis, A systematic literature review of blockchain-based applications: current status, classification and open issues, Telematics Inform. 36 (2019) 55–81, https://doi.org/10.1016/j.tele.2018.11.006.

[36] S. King, S. Nadal, Ppcoin: peer-to-peer crypto-currency with proof-of-stake, SelfPublished Paper 19 (2012). August.

[37] G. Greenspan, Ending the bitcoin vs blockchain debate, http://www.multichain.com/blog/2015/07/bitcoin-vs-blockchain-debate, 2015. (Accessed 10 March 2020).

[38] Hyperledger, Open source blockchain technologies, https://www.hyperledger.org/, 2020. (Accessed 20 March 2020).

[39] Bitshares, Your share in the decentralized exchange, https://bitshares.org/, 2020. (Accessed 18 March 2020).

[40] D. Schwartz, N. Youngs, A. Britto, The ripple protocol consensus algorithm, Ripple Labs Inc White Paper 5 (2014).

[41] K. Christidis, M. Devetsikiotis, Blockchains and smart contracts for the internet of things, IEEE Access 4 (2016) 2292–2303.

[42] D. Bhowmik, T. Feng, The multimedia blockchain: a distributed and tamper-proof media transaction framework, in: International Conference on Digital Signal Processing DSP, 2017.

[43] R. Xu, L. Zhang, H. Zhao, Y. Peng, Design of network media's digital rights management scheme based on blockchain technology, in: Proceedings of 2017 IEEE 13th International Symposium on Autonomous Decentralized Systems (ISADS 2017), 2017, pp. 128–133.

[44] J. Lin, Z. Shen, C. Miao, Using blockchain technology to build trust in sharing LoRaWAN IoT, in: ACM International Conference Proceeding Series, vol. Part F130655, 2017, pp. 38–43.

[45] N. Kshetri, Can blockchain strengthen the internet of things? IT Professional 19 (4) (2017) 68–72.

[46] K.R. Özyilmaz, A. Yurdakul, Work-in-progress: integrating low-power IoT devices to a blockchain-based infrastructure, in: Proceedings of the 13th ACM International Conference on Embedded Software 2017 Companion, EMSOFT, 2017, https://doi.org/10.1145/3125503.3125628.

[47] E. Mengelkamp, B. Notheisen, C. Beer, D. Dauer, C. Weinhardt, A blockchain-based smart grid: towards sustainable local energy markets, Comput. Sci. Res. Dev. 33 (1–2) (2018) 207–214.

[48] L. Wu, K. Meng, S. Xu, S.Q. Li, M. Ding, Y. Suo Democratic, Centralism: a hybrid blockchain architecture and its applications in energy internet, in: Proceedings – First IEEE International Conference on Energy Internet, 2017, pp. 176–181.

[49] X. Wang, L. Feng, H. Zhang, C. Lyu, L. Wang, Y. You, Human resource information management model based on blockchain technology, in: 2017 IEEE Symposium on Service-Oriented System Engineering (SOSE), San Francisco, CA, 2017, pp. 168–173.

[50] J. Wang, M. Li, Y. He, H. Li, K. Xiao, C. Wang, A blockchain based privacy-preserving incentive mechanism in crowdsensing applications, IEEE Access 6 (2018) 17545–17556.

Computational intelligent techniques for prediction of environmental attenuation of millimeter waves

Hitesh Singh[a], Vivek Kumar[a], Kumud Saxena[a], and Boncho Bonev[b]

[a]Noida Institute of Engineering and Technology, Greater Noida, India, [b]Technical University of Sofia, Sofia, Bulgaria

Chapter outline

Security and privacy issues in IoT devices and sensor networks. https://doi.org/10.1016/B978-0-12-821255-4.00012-2

1 Introduction

In the past decade, technologies related to higher-frequency bands, that is, millimeter waves, are on high demands. Different propagation studies have been done in the area of millimeter waves, but it is quite sensitive toward the existing conditions of the environment like rain, fog, oxygen, and dust. High-frequency waves that are above 10 GHz are sensitive toward environmental conditions. The impact of cloud on radio wave propagations over millimeter waves is significant, which can be observed from previous data.

So, working on high-frequency waves will produce a result that will be better as compared with other waves. The study needs to be conducted on the millimeter waves and to produce a model that will implement on high-frequency wave and show the effect of environment conditions. The model that will be developed on millimeter waves will have a more specific and accurate data on the effects of rain, fog, oxygen, dust, etc.

The chapter is organized in the following way. Section 1 describes about Earth's atmosphere and its composition. Section 2 is consisting of terrestrial and satellite links and impact of various atmospheric impairments of the atmosphere. Section 3 describes about the attenuation in radio waves due to the presence of clouds. Sections 4 and 5 discussed about the impact of rain on radio wave propagation for both satellite and terrestrial links. Section 6 shows the implementation of various rain models. Section 7 describes about the machine learning models reliance in the radio wave propagations for outdoor environmental conditions.

1.1 Atmosphere

Earth's atmosphere consists of different types of gases surrounding the globe that are dependent on each other due to the different revolution and rotation of the Earth. The atmosphere is in the form of concentric layers that determines different regions of atmosphere; homosphere and heterosphere are among these regions. Homosphere lies at the height of nearly 80 or 90 km, whereas above this, altitude extends heterosphere [1].

The layer of the atmosphere, which is nearest to the Earth, is called troposphere, which is characterized by the gradual decline of temperature ranges from −5 to −6°C per kilometer with the increase of altitude. The cloud formation, occurrence of rain, and various meteorological phenomena occur here. This layer can be categorized into two, namely, the free atmosphere and turbulent layer. The turbulent layer lies at the altitude of around 1500 m above surface and to higher altitudes where numerous interactions, either thermal or mechanical, between the surfaces of terrestrial environment and the atmosphere, are taking place. The effects of the temperature at the surface beyond this layer are negligible. The height of troposphere varies throughout the globe. For example, at the poles, it is 8 km high as compared with 18 km at the equator that depends on the geographical latitude including seasons and meteorological conditions. Excluding the cumulonimbus clouds, the variation in the temperature of the troposphere from 190 K to 220 K from the equator toward the poles, as the increasing convection movements of air slows down and forms an upper limit for the clouds.

1.2 Composition of atmosphere

With the change in altitude, contains of the atmosphere also vary. Gases become lighter with the rise of altitude. The constituents of the atmosphere are categorized into fixed density, variable density, and aerosols. A quasi-uniform distribution is observed in the most atmospheric constituents from 15 to 20 km. The nitrogen that are around 78.095% of the total volume is the most important followed by oxygen, argon, and carbon dioxide about 20.93%, 0.93%, and 0.03%, respectively, of total volume [1].

Water vapor is one of the main constituents. The concentration of water vapor strongly varies in atmosphere and depends on different geographical conditions. In the atmosphere, water presents in the form of solid ice crystals and snowflakes and in liquid form that is present in clouds, fog, and rain. These phases of water are responsible for different energy absorption effects like scattering of radio waves on the millimeter-wave frequencies [1].

Aerosols consist of fine particles whose size varies from 10 to 100 μm, with a low fall speed that is due to the effects of gravity. They can exist in liquid or solid form. As the dimensions of aerosols are very near to the wavelength of radio waves specially millimeter waves, they create disturbances in their propagation.

2 Terrestrial and satellite links

Radio relay links are also known as terrestrial fixed links, based on the radio wave propagation in troposphere where the significant meteorological phenomena occur like rain, snow, fog, hail, and clouds. The study of terrestrial fixed links comprises different phenomena of different nature, like reflection, attenuation, transmission, refraction, scattering, or depolarization phenomena. The electromagnetic waves above 10 GHz interact with the troposphere. Due to these interactions, energy is absorbed or diffused hence a change in the transmitted signals. This [1] shows that attenuation caused by water vapor is higher at frequencies greater than 22 GHz, while very less attenuation is caused by the presence of hail and crystals of ice.

In case of satellite links which are either operating between the Earth base station and a satellite or between satellite and Earth are characterized by the slanted toward the direction of propagated waves [1]. To study the impact of atmospheric impairments on propagation of radio waves between two base stations, it can be simplified by comparing the terrestrial links of the stations at the Earth. The ground reflection and diffraction phenomena can be eliminated as the influence of the meteorological phenomenon over the surface of ground is negligible. Atmospheric paths for the most part of the waves are neglected. Besides free-space attenuation, several factors are there by which it is necessary to study the radio wave propagation between Earth station and satellite.

2.1 Attenuation due to gas

Experimental studies show the effects of O_2 and H_2O vapor particles on radio waves for satellite communication at Ka-bands [2]. Different parameters like temperature, pressure, and relative humidity are observed from atmospheric infrared sounder, which is NASA's satellite. Signals coming from the satellite of Nigeria NigComsat-1 are observed from different stations located in the globe. It has been observed from the experimental study that attenuation caused by water vapor for Ka-bands is around 0.7–1.1 dB. Another observation found that attenuation caused by oxygen is higher at North West part of Nigeria as compared with other part of region.

It has been observed that in millimeter band some atmospheric gases like O_2, NO_2, SO_2, H_2O, and N_2O have absorption lines, though the water vapor and oxygen molecule are the main causes [3]. As other gases have low density, the loss in absorption is negligible.

Two absorption lines of water vapor that is an electric dipole polar molecule are seen in the millimeter-wave band at 22.235 GHz and 183.31 GHz and in the infrared field another absorption lines above 300 GHz. For frequencies below 100 GHz where absorption lines are negligible, the coefficient of the total water vapor absorption can be expressed as [3] (in dB/km)

$$k_{H_2O} = 2f^2 p_v \left(\frac{300}{T}\right)^{1.5} \gamma_1 \left[\left(\left(\frac{300}{T}\right)e^{\frac{-644}{T}}G + 1.2 \times 10^{-6}\right)\right] \tag{1}$$

where

k_{H_2O} = vapor absorption of water (H_2o)
f = frequency of waves in GHz
T = atmospheric temperature measured in Kelvin
p_v = content of water vapor in g/m^3
atmospheric pressure in mill bars

$$G = \frac{1}{(494.4 - f^2)^2 + 4f^2\gamma_1^2} \tag{2}$$

$$\gamma_1 = \left(\frac{P}{1013}\right)\left(\frac{300}{T}\right)^{0.626}\left(1 + 0.018\frac{p_v T}{P}\right) \tag{3}$$

The water vapor content at sea level can differ from 0.001 g/m^3 to 30 g/m^3 depending on cold, dry climate to hot, and humid climates. At the mid-latitudes, around 45 degrees N, average water vapor [3] content is 7.72 g/m^3.

Due to the presence of permanent magnetic moment in oxygen molecule, the absorption lines are observed from 50 GHz to 70 GHz [3].

In case of lower atmosphere, the absorption lines lie around 60 GHz and 118.75 GHz. The f_0 represents the absorption line at 60 GHz:

$$k_{O_2} = 0.011 f^2 \left(\frac{P}{1013}\right)\left(\frac{300}{T}\right)^2 \gamma \left(\frac{1}{\left(f - f_o^2\right)^2 + \gamma^2} + \frac{1}{f^2 + \gamma^2}\right) \tag{4}$$

where

k_{O_2} = vapor absorption of oxygen (O_2)

$$\gamma = \gamma_0 \left(\frac{P}{1013}\right)\left(\frac{300}{T}\right)^{0.85} \tag{5}$$

$$\gamma_0 = \begin{cases} 0.59 & P \geq 333\,\text{mbar} \\ 0.59[1 + 0.0031(333 - P)]25\,\text{mbar} \geq & P \geq 333\,\text{mbar} \\ 1.18 & P \leq 25\,\text{mbar} \end{cases} \tag{6}$$

2.2 Attenuation caused by snow

Snow that is in the form of ice crystal or flakes has a diameter between 2 mm and 5 mm. Sometimes, it may rise to 15 mm also. The structure is a bit complicated in nature. The measurements carried out through photographs revealed that there was a difference between maximal horizontal and vertical dimensions [4]. Based on the flake diameter, the average value of this ratio varies from unit to 0.9 mm. The experimental study shows that for large densities of snowflakes, a semiempirical model for the terminal fall speed of snowflakes [4] is valid:

$$v(D) = 3.94\sqrt{\frac{D(\rho_s - \rho_a)}{2}} \tag{7}$$

where

$v(D)$ = terminal velocity (m/s)
ρ_s = density of snowflakes (g/cm^3)
ρ_a = air density in (g/cm^3)
D = diameter of snowflakes (mm)

Dimensions of snowflakes are determined by Gunn distribution [5]. In this experimentation, two parameters are used, which is given by N_0 and Λ that are in mm^{-1} m^{-3} and cm^{-1} [5]:

$$N_0 = 3.8 \times 10^3 R^{-0.87} \tag{8}$$

$$\Lambda = 25.5 R^{-0.48} \tag{9}$$

where R (mm/h) is the rate of snow fall [6].

This shows that this model is mostly designed to snowflakes with relatively small water contents.

And the parameters are

$$N_0 = 2.5 \times 10^3 R^{-0.94} \tag{10}$$

$$\Lambda = 22.9 R^{-0.45} \tag{11}$$

2.3 Attenuation due to hail

At very low temperatures, precipitation of ice particle occurs inside cumulonimbus clouds called as hail. Very few studies have been conducted for this phenomenon as it has a much localized character. In the hail the ice contains air bubble that forms hollow structure that is the largest observable hydrometeors. There are different forms of hail stones like spheroid, spherical, conical, and other irregular forms. In case of small hailstones, the most common form used is spherical form. The theory of diffusion by two concentric spheres with different refractive indexes by Oguchi [7] explains the scattering properties of hail at millimeter radio wave and the concept of the terminal fall velocity of hailstones studied in Douglas [8]. Models by Douglas and Smith explain the distribution of hailstones based on size [8]:

Douglas:

$$N(D) = 4960 \times R^{-1236D} \tag{12}$$

Smith (weak hail, $R = 10$ mm/h):

$$N(D) = 1.1 \times 10^5 \times R^{-2000D} \tag{13}$$

Smith (strong hail, $R = 100$ mm/h):

$$N(D) = 5.8 \times 10^4 \times R^{-1080D} \tag{14}$$

where R is precipitation rate of hailstone after fusion.

2.4 Attenuation due to dust

The studies show the effects of dust and sandstorms on satellite communication for Ka-bands [9–11]. This results in a new dust storm-based model for the identification of visibility, fragmenting the visibility data based on the height and presenting the three-dimensional relationship of radio wave attenuation with visibility, propagation angle, frequency, and dust particle size. The results obtained with the simulation are compared with the experimental results obtained from VSAT and DVB-32 satellite systems.

2.5 Attenuation due to scintillation

Mohammad et al. studied the effect of scintillation in tropospheric layer over the region of Bangladesh [12], and they have done the experimental study by using ITU scintillation model for prediction of attenuation. It has been observed that scintillation fade depth (SFD) mostly affects the region Rajshahi followed by Chittagong. For regions Dhaka and Sylhet, it shows that a value of SDF varies from 3 dB to 13 dB, respectively.

Another study is done in the region of Madrid to measure the effect of scintillation on Ka-bands by Pedro Garcia del Pino et al. [13]. Deviation of nearly 4 dB was observed for the hot seasons. An observation has been made from experimentation that there is a relationship between intensity and temperature of cumulus clouds.

Different models proposed till now for predicting the attenuation caused by scintillation like Allnutt, Karasawa, Yamada, and ITU-R have been discovered by using measurements of different satellites of the United States, Europe, and Japan [14]. From the comparative study, seasonal variation of scintillation intensity was observed. An observation has been made that scintillation was related to cloud turbulence.

3 Cloud attenuation

A cloud consists of mass of small droplets of water, super cool water, ice crystals, or particles suspended in air. Saturation of atmospheric air is responsible for cloud formation. Another reason for the formation of clouds is due to the presence of sufficient moisture that rises to ambient temperature.

Literature has shown that in troposphere there are several types of clouds. These are divided into three main categories, namely, low-level clouds, that is, cumulonimbus, cumulus, stratus, and stratocumulus; middle-level clouds including altocumulus, nimbostratus, and altostratus; and high-level clouds, that is, cirrocumulus, cirrus, and cirrostratus [15].

3.1 Cloud attenuation model

In the past decade, various studies have been conducted for cloud attenuation. In this section, famous models are discussed.

Theoretical results have been described in the form of equations and figures [16]. Results obtained from mathematical equation are compared with observations of other authors. The input parameters of the proposed system include water density (g/cm), liquid water content (gm/m^3), and K that is the imaginary part of absorption coefficient and wavelength (cm).

LWC of clouds and distribution of cloud at different heights of clouds are presented by Staelin [17]. There is a difference between the accuracy for LWC and accuracy for water vapor distribution. The result of liquid water content is less.

In Slobin's work [18] a general overview about types of cloud, density of liquid water, and noise temperature and radio wave attenuation that is caused by cloud calculations is shown. Experimental studies for 15 different locations of America have been done. The attenuation is calculated due to cloud and noise temperature. Tests are performed for all calendar years, and for experimentation, more precise data are required. In this work, data for elevation angle are missing.

A study has been done for predicting the attenuation caused by clouds at the frequency ranges from 15 GHz to 35 GHz at the site of Boston, USA [19]. Data for partial and full cloud cover are collected. From the experimentation, it has been observed that attenuation due to clouds is proportional to the slant path distance. In this work, relation between attenuation and humidity has been derived. It has also been observed that attenuation caused by full cloud cover is significant as compared with attenuation observed for clear sky.

Experiment was performed to calculate the attenuation caused by rain, clouds, haze, fog, dry air, and water vapor by Liebe [20] for frequency ranges up to 1000 GHz. Based on these readings, mathematical model has been derived. The input parameters used by them are temperature, rainfall rate, suspended water droplets, pressure, and relative humidity.

Another model has been developed to calculate the attenuation caused by clouds for satellite communication link [21]. Basic parameters used in this model are temperature and humidity. The experimentation has been performed for three different locations of Europe. For satellite link the elevation angle of 15–40 degrees has been used.

An empirical model has been proposed to calculate fading of angle and attenuation due to clouds and rain [22]. This model works effectively for 10-degree elevation angle. From the experimentation performed on beacon and radiometer, data for slant path length have been calculated.

A semiempirical model to estimate cloud attenuation has been proposed [23]. They have collected the data for surface temperature and humidity. It was the modification of Sobin's model [18] for rain attenuation caused by cloud. Experimentation has been carried out at different frequencies between 20 GHz and 30 GHz.

Another method has been described to predict aircraft-aircraft, satellite-Earth, and aircraft-Earth link attenuation [24]. Satisfactory results were obtained in this method, and good prediction has been given for attenuation caused by cloud.

Another method has been proposed for estimating the attenuation caused by both clouds and gases [25]. This method has been effectively working for millimeter waves. For the experimentation the satellite used is ITALSAT F1, which is of Italian Space Agency.

Most widely used model by engineers and researchers is the ITU-R model [26]. This model is designed after the huge amount of data collected and analyses thought the world. This model consists of curves, graphs, and equations.

3.2 **Work done by other researchers**

In the earlier section, various researchers have proposed mathematical models for prediction of radio wave attenuation caused by clouds. These models are based on mathematical formulations and statically analysis. Now we are going to discuss about the propagation studies done by other researchers in different parts of globe.

To identify the accuracy of water vapor content and liquid water content of clouds, climatological data have been observed [27]. It has been observed that there are certain uncertainties that are observed while calculating water vapor absorption. A channel model to minimize these uncertainties has been designed. It has also been observed that the attenuation effects of cloud are reduced if the effecting temperature radiated from clouds is known. To calculate this temperature a separate reading has to be taken for base height and cloud thickness.

Most researches are performed for clouds, but very few have done for the ice contents of clouds. Papatsoris [28] has done the propagation studies about ice content of clouds. They have done experimentation based on cross and linear polarization for two different scenarios. The first scenario was based on aircraft-to-aircraft ink in which cross polarization discrimination was observed. In this scenario the polarization angle was 20 degrees, and ice column was 30% aligned. The second scenario was intersatellite links. In this scenario the polarization angle was 10- with 3-degree elevation angle. And ice columns are aligned by 50%.

A cloud attenuation model has been presented by Dissanayake et al. [29]. It was based on the occurrence probability of four different types of cloud. The probability of occurrence of cloud is derived from various meteorological stations present throughout the world. Results obtained from this model were compared with the data from cloud attenuation observed by satellite beacon and radiometer.

A research has been carried out to study the occurrence of clouds at different parts of Indian subcontinent [30]. They have observed a deduced occurrence of clouds for low-level cloud coverage for the region of Hyderabad, India. Data are observed for different months and at daytime and nighttime. They have done experimentation for frequencies ranges from 10 GHz to 100 GHz. They have observed that for liquid water content of about 1 g/m^3, the attenuation at 30 GHz was 0.88 dB/km and for 75 GHz is about 5.55 dB/km.

A cloud attenuation study mainly focused on eastern part of India has been performed [31]. The cloud occurrence data for different months and time are observed. The thickness of clouds is also observed. The experimental study has been conducted at frequencies 10 GHz, 18 GHz, 32 GHz, 44 GHz, and 70 GHz, respectively.

An experiment for prediction of attenuation caused by clouds and rain is performed [32]. The cumulative results of their detailed study were presented graphically. They have done the study from 2000 to 2007 at different parts of India.

An experiment has been carried out for study of attenuation caused by clouds for Penang region by using SUPERBIRD-C satellite beacon [33]. The occurrence of cloud is observed throughout the year. Five-year data have studied for cloud attenuation from satellite and then make analysis of measured data in different proposed model.

Animesh et al. [34] have carried out measurements at Kolkata region of India by using the results obtained by implementation of Salonen model [21] and compared it with ITU-R model. They have also observed the occurrence of cloud throughout the year and collected liquid water content from radiosonde data. They have done measurements for frequencies from 10 GHz to 100 GHz. They have observed that cloud attenuation at frequency below 50 GHz is lesser as compared with the attenuation observed in ITU-R model.

Mattioli et al. [35] analyzed the performance of the Decker and Salonen for non-precipitation cloud attenuation, and based on their results a new cloud model was predicted. A new cloud density parameter was introduced for liquid water content of cloud and ice content. The performance of this parameter was examined at Atmospheric Radiation Measurement Program of the United States. This new parameter has shown improved results in calculating cloud detection about 15%. The overall results of the simulated model are compared with measured data, and satisfactory measurements are observed.

Temidago et al. [36] have done long-term measurements from 1953 to 2011 for different parameters of clouds at six climate zones of Africa. The presented results show good correlation between monthly distribution of cloud temperature, cloud cover, cloud top height, and precipitation frequency at different six zones of Africa. The liquid water content of cloud observed for six zones shows degradation from the liquid water content of ITU-R model from 32% to 90% occurring of 0.01% to 10% of time. It has observed that cloud attenuation at tropical rain forest is high because of greater occurrence of rain and different cloud parametric factors.

Bijoy [37] has studied attenuation caused by cloud ice content and attenuation caused by cloud liquid water content for different millimeter-wave frequencies at different tropical regions. The results obtained from experimentation show that attenuation caused by liquid water content is higher than attenuation caused by ice content in troposphere.

Ahmed Ali et al. [38] have done experimental study for cloud attenuation at different sites for Sudan. They have observed that low cloud occurrence plays a significant role for attenuation. According to the author's observation, cloud attenuation and cloud noise temperature play a significant role at radio wave attenuation caused by clouds. They have done experimentation at 10 GHz, 16 GHz, 32 GHz, 44 GHz, and 70 GHz.

4 Rain attenuation

Various studies on rain attenuation have shown that the distribution of rain is inhomogeneous along the path of radio waves. The variations of rainfall in both vertical and horizontal directions make the attenuation caused by slant path more complex. It has been observed that attenuations due to rain and gases at higher frequencies are significant. The heavy rainfall is usually confined to smaller areas and has shorter span of time as compared with lighter rain. The uncertainties of rainfall rates and rain cells size may lead to incorrect estimation of rain attenuation.

Ka-band [39] is being used for advanced communication satellite systems to provide data with high throughput to smaller terminals, thus catering to the needs of the time. However, as these high-frequency Ka-band signals are strongly impaired by various impediments, especially the attenuation due to rain is the most severe degradation. Rain attenuation leads to degradation in the received strength (C/No) of a signal and in turn deteriorates the performance of the system. Therefore it is necessary to estimate the expected rain attenuations accurately to design a satellite link in this band. To study the impairment in rain attenuation, six different ground stations all over India have been set up across India that measures the rain rate and drop size distribution (DSD) along with the received signal level transmitted from a 20.2 and 30.5 GHz beacon, currently on board GSAT-14 satellite.

Reliable modeling [40] of atmospheric attenuation is needed for optimum design of satellite communication links. In the view of upcoming satellite communications in this region, it is necessary to study the atmospheric attenuation at Ka-band during rain events. In this paper, attenuation estimation at Ka-band obtained from radiometric brightness temperatures (Tb) at Kolkata, India (22034′E, 88022′N), has been presented. The contribution of water vapor is examined in determining the behavior of attenuation during rain at frequencies 22.24 GHz, 26.24 GHz, and 31.4 GHz. It is observed that though the attenuation at 31.4 GHz is the highest, the attenuation at 22.24 GHz is higher than that at 26.24 GHz due to water vapor absorption under low-to-moderate raining conditions. The attenuation data obtained from ITU-R and millimeter-wave propagation (MPM) model are compared with the radiometric data to understand the limitation of radiometric measurements.

The effects of rain on millimeter-wave communication are investigated in the present study [41]. Propagation of very short pulses at millimeter-wave band has been discussed. Specific attenuation has been computed at different millimeter-wave frequencies, and its variation as a function of rain rate and frequencies has been assessed. The variation of rain attenuation over terrestrial link as a function of path length, frequency, and rain rate has been examined. A comparison between rain attenuation along terrestrial path and earth space path has been made at millimeter-wave band.

ITU-R rain model has been developed by the International Telecommunication Union for the prediction of rain attenuation based on the probability of rain rate in the percentage of time.

4.1 ITU-R model

The recommendation ITU-R Rec. P. 618-10 rain attenuation model is the widely used method for the predictions of rain effects on satellite communication system [42].

The attenuation is calculated for 99.99% fade depth by

$$\text{Att}_{0.01} = kR^{\alpha}dr \ dB \tag{15}$$

where R is given by 99.99% of rain rate for rain region in mm/h, kR^{α} is the specific attenuation in dB/km, and d is the link distance in kilometer. The formula for r is given as

$$r = 1/(1 + d/d_0) \tag{16}$$

where

$$d_0 = 35e^{-0.015R} \ \text{km} \tag{17}$$

where d_0 is the effective path length and r is called the distance factor. K and α are the regression coefficient for frequencies and polarization [4].

4.2 Simple attenuation model

To predict the attenuation caused by rain, simple attenuation model [42] has been proposed. It was effectively based on 10–35 GHz. The attenuation due to rain can be calculated by

$$A = \gamma \frac{1 - e\left[-\gamma bln\left(\frac{R_{\%p}}{10}\right)\right]L_s Cos\theta}{\gamma bln\left(\frac{R_{\%p}}{10}\right)cos\theta} R_{\%p} > 10 \ \text{mm/h} \tag{18}$$

where L_s is called slant path in kilometer and θ is elevation angle in degree that was between the slant path and horizontal projection. Specific attenuation is calculated by the formula $\gamma = k(R_{0.01})^{\alpha}$ (dB/km), and empirical constant is given by $b = 1/22$.

Following expression for effective rain height H_R based on calculated data can be obtained as

$$H_R = \begin{cases} H_0; & R \le 10 \ \text{mm/h} \\ H_0 + \log\left(\frac{R}{10}\right); & R > 10 \ \text{mm/h} \end{cases} \tag{19}$$

Calculation from given formulation generates a function of average rain rate for rain attenuation and with the combination of those with rain statistics can estimate different statistics of rain attenuation annually.

4.3 García-López method

A simple method was developed by García-López et al. [42] by inserting different parameters. Model shows good results regarding rain attenuation. This model calculated the values of rain attenuation by using different coefficients for tropical region. The proposed method by García-López et al. [39] to calculate rain attenuation can be calculated by

$$A = \frac{kR^{\alpha}L_S}{\left[a + \left\{\frac{L_S(bR + CL_S + d)}{e}\right\}\right]} \tag{20}$$

where R denotes the amount of rainfall in mm/h; k and α are the coefficients based on parameters like frequency, polarization, and elevation angle; a, b, c, d, and e are constant values given by the model; and L_S (km) is the slant path up to rain height. In expression (20), e is a scaling factor of rain attenuation and, by taking $e = 10^4$, worldwide coefficients:

$a = 0.7, b = 18.35, c = -16.51$, and $d = 500$ (based on geographical area). For tropical climates, $a = 0.72, b = 7.6, c = -4.75$, and $d = 2408$. In this model the rain height H_R is given by

$$H_R(km) = \begin{cases} 4, 0 < |\varphi| < 36^o \\ 4 - 0.075(|\varphi| - 36^o \end{cases} \tag{21}$$

where φ denotes latitude of the earth station and is calculated in degrees.

4.4 Model proposed by Brazil

This model was developed for different rain distribution. Attenuation is calculated by

$$A_p = k\left[1.763\,R^{0.133 + 0.197/L_S\cos\theta}\right]^{\alpha} \frac{d}{1 + \frac{d}{119R^{-0.244}}} \tag{22}$$

where elevation angle is denoted by θ.

4.5 RAL model

A proposed model is given by Rutherford Appleton Laboratory (RAL) for rainfall rate measurements, which is used to develop a new model to demonstrate number of events of given time frame with rainfall having specific thresholds. They comprise rain rate measurement of 3-year data using three rapid response rain gauge and space size 200 m apart. All the measurements were carried out at Chilbolton, Hampshire (ITR-U), having sampled at 10-s intervals. The power-law model and lognormal model were consider in this approach, where R denotes rain rate exceeding and t_d denotes duration exceeding.

N denotes number of rain events per year, which is calculated by

$$N = 1.70 \times 10^4 R^{-1.76} e\left\{-\frac{(\ln t_d - 2)^2}{3.86 - 0.0409R}\right\} \tag{23}$$

Different models discussed earlier are used for the prediction of rain attenuation like crane, ITU-R, Brazil, and DAH. Comparative study of implementation of some rain attenuation models like ITU, Gracia, RAL, and Brazil is presented in Fig. 1.

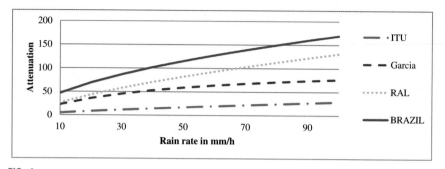

FIG. 1

Comparison of various rain models.

5 Rain attenuation in terrestrial links

The work done by Kesavan et al. [43] describes the overall comparative study of various other attenuation models proposed by researchers especially for tropical zone. The models that are studied are the revised Moupfouma, ITU-R, revised Silva Mello, and Lin model. Results from experiments have shown that they are useful for researchers and engineers in decision making for predicting the attenuation caused by rain at different zones of tropical environment. It has also been observed that results obtained from the model proposed by ITU-R are underestimated for higher frequencies. The results obtained from experiments show that the proposed model for predicting attenuation caused by rain at tropical region of Malaysia has shown good results compared with other geographical regions.

Another research has been carried out by Tamošiūnaitė et al. [44] for the region of Lithuania. The relationship obtained from experimentation between the rain rate date of 1-min data and the real-time values of rain rate observed for longer duration is used. The values obtained from real-time rain rate data for 1 min for the region of Lithuania are observed from the weather station situated at this region. The real-time value of rain rate observed from the weather station is compared with the simulated values obtained from proposed model and ITU-R model. By using the observed values of rain rate, a new method for calculating 1-min rain rate is derived. By using the earlier experimentation, the lowest value of rain rate for the region was also observed. It has also been observed that the mathematical calculation of specific attenuation caused by rain by using the real-time data is two times higher than the attenuation calculated using ITU-R model.

In the work done by Folasade Abiola Semire et al. [45], 2-year rainfall rate data at the region of Ogbomosho station have been used to study the integration time effect on cumulative distribution of rain rate. Experimental result obtained from experimentation shows that there is a relationship between the equiprobable rain rates for two different integration times and the conversion factors are climate and terrain dependent. The conversion factors CR and CE observed at Ogbomoso are lower as compared with results observed by Ile-Ife. This is because of the fact that there are differences in rain gauge sampling frequency, sensitivity, and accuracy; regional rain rate differences; topography; and climatic conditions and the effect of global warming. It has been observed that different factors of conversion are required for different scenarios even for the same climatic zones for the conversion from one integration time to another as against the ITU-R unified time integration regression coefficients.

An analytical study has been carried out for the measurement of distributed fade slop that was obtained from rain attenuation values, measured at the site of Eindhoven University of Technology from the satellite Olympus Max M. J. L. van de Kamp [46]. The measurements show that the fade slope distribution for both negative and positive values is similar and independent for frequency ranges from 12 GHz to 30 GHz and used the standard deviation parameter for calculating conditional distribution. It has been found that this parameter is directly prepositional to attenuation and depends on the type of rainfall. The relationship observed between attenuation and standard deviation is compared with observation measured from another site. From the results, it has been observed that fade slope deviation depends on elevation angles and on climatic conditions.

In the experimental study done by Obiyemi et al. [47], conversion coefficients are obtained for predicting the equivalent 1-min rain rate statistics available from 5 to 30 min over Akure, South-Western Nigeria. The mathematical derivation depends on the power-law function existing between equiprobable rain rate statistics, while comparison is made with estimates available for different locations. Apparent in the result is the significant variation in the values estimated for similar observation periods over the years and across different locations. From the observation of the coefficients derived for Akure, the 5-min conversion provides more satisfactory results and hence is more suitable for predicting the 1-min rain rate for the region of Akure.

The drop size distributions (DSD) of rain are calculated with the help of an instrument called disdrometers at various climatic zones of Indian subcontinents [48]. It has been assumed that the long normal DSD that is drop size distribution is used to derive the model of attenuation caused by rain at the frequency range from 10 GHz to 100 GHz. Attenuation caused by rain is calculated by assuming that a scattering is performed on single spherical rain drops. DSD depends on different climatic conditions in which different characteristics of attenuation are observed. Another observation indicates that significant difference between DSD-derived values and values obtained from ITU-R model at high frequency is measured for high rain rates at different climatic zones. The results observed from the experimental studies will be used to understand the different patterns of rain.

The work done by Kesavan et al. [49] shows the preliminary results obtained from effects of rain at 23-GHz point-to-point terrestrial 1.3-km-long communication link. The experimental setup was deployed in the region of Malaysia, which is the equatorial climatic zone of monsoon having maximum rainfall rate of about 100 mm/h. To maintain the effective performance of radio link, mitigation technique has to be implemented. A solution that was based on frequency variation has been proposed in this work. Another radio link was placed parallel to the main link that was based on frequency ranges not affected by the rain. This experimentation provides the effective solution for the outage condition due to the different rain rates.

The parabolic equation model for predicting radio wave attenuation at millimeter waves that was based on the effective permittivity is developed by Nen Sheng et al. [50], by modifying the refractive index. Finally the model is simulating the propagation characteristics of millimeter wave in different geographical regions, which have irregular terrain and sea surface that are rough in nature, and in complex environmental conditions of standard atmosphere conditions like rain and fog.

In the work by [51], author has observed the effects of rain on radio wave propagation. The main aim was to observe the localized behaviors of rain and its effects on millimeter-wave propagation. A millimeter-wave model called Tokyo Tech network was used on the measurement of signals and rainfall at the university campus. It has been observed that the localized behaviors of rain affect the rain attenuation at both rain rates and path distance.

To design a satellite communication link for the frequency ranges of above 10 GHz, 1-min rain rate statistics are important [52]. There is a need for the prediction model for attenuation caused by rain to overcome the degradation in radio signal. A model has been proposed by analyzing the rain statistics of 3 years at equatorial. A comparative study has also been performed between rain rate statistics of 1 min and predicted rain model.

Experimental study shows that different parameters for calculating DSDs affect the radio wave propagation [53]. Regression coefficients k and α in specific rain attenuation formula $\gamma = kR^{\alpha}$ where R is rain rate in mm/h which is derived from the experimentation. This experiment work is performed at England and Singapore. From the experimentation, it has observed that drop size distribution differs significantly at given two sites.

Experimentation for 60-GHz millimeter wave has been done [54]. The purpose of this experimentation is to find the relationship between rainfall rate and measured specific attenuation of the radio wave signal. A detailed statistical analysis has been done on experimental data, and the results are compared with the ITU-R model. The rain rate was measured at different sites, and observed results are matched with ITU-R model. It has been observed from the experimental study that observed results were higher than results obtained from ITU-R model.

An overview of the ITU recommendations that was based on error and performance objectives is described in the work by [55]. In this work, both the fade margin and link power budget observations are presented. Rain rate characteristics along with the rain attenuation models are presented. Statistics of attenuation due to rain that are needed for the availability performance assessment are also introduced monthly and yearly. In this work the experimental setup of radio link at 38 GHz, 58 GHz, and 93 GHz has been implemented. The experimentations are performed at the geographical regions of Prague, the Czech Republic. The results obtained from the experimentation show both rain attenuation and rain intensities. The availability performances based on experimental results are also been assessed. In this study, both the polarization scaling method and ITU-R frequency scaling method and proposed path length scaling method are presented.

An improved version of the EXCELL has been proposed to predict attenuation due to rain [56]. With the help of this improved rain model, the prediction of rain attenuation from both convective and stratiform rain can be calculated. The main objective of this work is the attempt to develop a more accurate model for compression with EXCELL model. In this experimentation, different rain heights, which were obtained from the ERA-15 database, are used for convective and stratiform rain, respectively. Moreover, in the case of stratiform rain, the bright band contribution to attenuation is added.

Another experimental study for 50 GHz shows that there is a relationship exists between attenuation caused by rain and observed scintillation was performed slant path attenuation [57]. This study was based on the probability density function (PDF) for different parameters of scintillation. It was observed that intensity of scintillation is calculated as the standard deviation of the amplitude fluctuations that was proportional to rain attenuation. It has also been observed that a methodology was described to mathematically derive the overall PDF of scintillation amplitude by using a combination of conditional PDFs and local data of rain attenuation.

For prediction of attenuation caused by rain, a new model was presented in this work [58]. The proposed model was based on Mie theory of electromagnetic waves scattering by dielectric sphere of raindrops, by assuming that the spherical nature of raindrops. Forward scattering amplitudes of the raindrops that were of spherical shape are presented, and the extinction cross sections are also been calculated. The power-law regression formulation has been applied to calculate the cross-sectional power-law coefficients. The attenuation caused by rain is modeled by integrating the extinction power-law model over different DSD models. The experimental setup has been done in Durban, South Africa, in which the attenuation was observed at 19.5-GHz frequency band in 6.73-Km-long propagation path. The results obtained from experimentation were compared with the mathematical model by using same propagation parameters.

6 Implementation results of rain model

Various models used for the prediction of attenuation caused by rain are discussed in Section 4. Those discussed models are implemented to do comparisons. The parameters used in this model are rain rate from 10 mm/h to 100 mm/h; frequency used is 60 GHz. Other parameters depend on the model.

It has been observed from the implemented results shown in Fig. 1 that Brazil model shows higher attenuation as compared with the ITU-R model. ITU-R and Garcia models show quite similarity with each other. It has also been observed that with the increase in rain rate, attenuation is also increased. At very little rain that is about 10 mm/h, the attenuation observed is around 3 dB for ITU-R model and 20 dB for

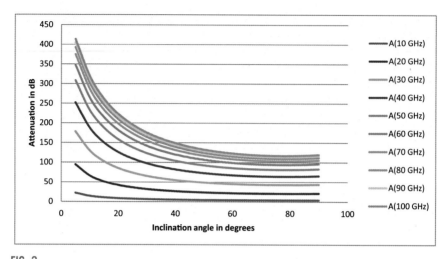

FIG. 2

Attenuation versus angle of inclination for different frequencies.

Gracia model, whereas for higher rain rate, attenuation of about 20 dB is obtained for ITU-R model and 70 dB for Gracia model.

To understand the overall impact of rain in the satellite link, it is important to study the radio wave propagation. The ITU-R [59] model for rain is the widely acceptable model to study these effects. The impact of rain has been studied for different frequencies from 10 GHz to 100 GHz at different inclination angles from 10 degrees to the 90 degrees. The extermination has been performed at the site of Indian subcontinent. Different parameters used in study are latitude that is 25.17 degrees and rain rate value that is 42 mm/h (Fig. 2).

From this experimentation, it has been observed that with low elevation angle about 10 degrees, the higher attenuation is observed. This is because of the fact that radio waves cover longer distances and more obstructions come on the way. It has also been observed that attenuation becomes constant after inclination of 60 degrees. The experiment has shown that with higher-frequency bands, more attenuation is observed, for example, at frequency about 60 GHz at 60 degree inclination, 100 dB of total attenuation is observed.

7 Issues related to machine learning

In the work done by [60], the researchers have developed and implemented a model that was based on fuzzy logic algorithm used for hydrometeor particle identification that is easy and efficient enough to implement in a real-time environmental condition. It has also been observed that there are not much detailed measurements available for this type of study. The results obtained from this work are encouraging. Plans for further optimization and verification of the proposed algorithm are described.

OCEAN is the CSI prediction scheme in which channel state information was predicted using historical data for next-generation wireless communication [61] and finds out affecting parameters for CSI radio links. Convolution neural network with memory network techniques was implemented. In the next-generation wireless communication, two-step training system has been implemented. To validate OCEAN's efficacy, different indoor and outdoor experimental studies have been done. The experimental results show that OCEAN not only gets the predicted CSI values quickly but also achieves more accurate CSI prediction with up to 2.650%–3.457% average difference ratio between the measured CSI and predicted.

Method for predicting the attenuation caused by rain for satellite link at Ka-band has been proposed [62]. In the experimental setup, there are millions of ground stations presented around the world. In this setup, prediction of attenuation due to rain is not feasible. To do prediction a neural network techniques are used in the proposed model. The implementation results of proposed model are matching with the proximity of 85% with the results obtained by rain gauge.

8 Conclusion

The impact of environmental conditions like rain, cloud, gases, and dust has significant impact on the radio wave propagation. It has observed that the effect of rain on electromagnetic waves is much more than the clouds. Different mathematical models have been discussed in this work, and they give predictions of attenuation caused by rain in the higher-frequency ranges. It has also been observed that machine earning methodologies are very effective for the prediction of attenuation due to hydrometeors. By using deep learning methodologies, it is very effective and efficient way to predict the results and helpful for the researchers to support highly sensitive future technologies. It is observed that if radio wave system uses deep learning methods, then attenuation problems caused by rain, cloud, etc. can be reduced.

References

[1] J.S. Seybold, Introduction to RF Propagation, John Wiley & Sons, 2005.
[2] T.V. Omotosho, J.S. Mandeep, M. Abdullah, Atmospheric gas impact on fixed satellite communication link a study of its effects at Ku, Ka and V bands in Nigeria, in: 2011 IEEE International Conference on Space Science and Communication (IconSpace), IEEE, 2011.
[3] T.L. Frey, The effects of the atmosphere and weather on the performance of a mm-wave communication link, Appl. Microw. Wirel. 11 (1999) 76–81.
[4] C. Magono, T. Nakamura, Aerodynamic studies of falling snowflakes, J. Meteorol. Soc. Jpn. Ser. II 43 (3) (1965) 139–147.

[5] K.L.S. Gunn, J.S. Marshall, The distribution with size of aggregate snowflakes, J. Meteorol. 15 (1958) 452–461.

[6] R.S. Sekhon, R.C. Srivastava, Snow size spectra and radar reflectivity, J. Atmos. Sci. 27 (2) (1970) 299–307.

[7] T. Oguchi, Electromagnetic wave propagation and scattering in rain and other hydrometeors, Proc. IEEE 71 (1983) 9.

[8] R.H. Douglas, Hail size distributions of Alberta hail samples, Mc Gill Univ. Montreal Stormy Wea. Gp. Sci. Rep. MW 36 (1963) 55–71.

[9] K. Harb, et al., A proposed method for dust and sand storms effect on satellite communication networks, in: Innovations on Communication Theory INCT (Istanbul, Turkey), 2012, pp. 33–37.

[10] K. Harb, et al., Systems adaptation for satellite signal under dust, sand and gaseous attenuations, J. Wirel. Netw. Commun. 3 (3) (2013) 39–49.

[11] K. Harb, et al., Ka-band VSAT system models under measured DUSA attenuation, in: SPACOMM, the Seventh International Conference in Advances in Satellite and Space Communications, 2015.

[12] S.M. Hossain, A.M. Samad, The tropospheric scintillation prediction of earth-to-satellite link for Bangladeshi climatic condition, Serbian J. Electr. Eng. 12 (3) (2015) 263–273.

[13] D. Pino, P. Garcia, et al., Tropospheric scintillation measurements on a Ka-band satellite link in Madrid, URSI (2008).

[14] M.M.J.L. van de Kamp, et al., Improved models for long-term prediction of tropospheric scintillation on slant paths, IEEE Trans. Antennas Propag. 47 (2) (1999) 249–260.

[15] World Meteorological Organization, Cirrus, International Cloud Atlus, 1975.

[16] K.L.S. Gunn, T.W.R. East, The microwave properties of precipitation particles, Quart. J. Roy. Meteorol. Soc. 80 (346) (1954) 522–545.

[17] D.H. Staelin, Measurements and interpretation of the microwave spectrum of the terrestrial atmosphere near 1-centimeter wavelength, J. Geophys. Res. 71 (12) (1966) 2875–2881.

[18] S.D. Slobin, Microwave noise temperature and attenuation of clouds: statistics of these effects at various sites in the United States, Alaska, and Hawaii, Radio Sci. 17 (6) (1982) 1443–1454.

[19] E.E. Altshuler, R.A. Marr, Cloud attenuation at millimeter wavelengths, IEEE Trans. Antennas Propag. 37 (11) (1989) 1473–1479.

[20] H.J. Liebe, MPM—an atmospheric millimeter-wave propagation model, Int. J. Infrared Millim. Waves 10 (6) (1989) 631–650.

[21] E. Salonen, S. Uppala, New prediction method of cloud attenuation, Electron. Lett. 27 (12) (1991) 1106–1108.

[22] A. Dissanayake, J. Allnutt, F. Haidara, A prediction model that combines rain attenuation and other propagation impairments along earth-satellite paths, IEEE Trans. Antennas Propag. 45 (10) (1997) 1546–1558.

[23] F. Dintelmann, G. Ortgies, Semiempirical model for cloud attenuation prediction, Electron. Lett. 25 (22) (1989) 1487–1488.

[24] T. Konefal, et al., Prediction of monthly and annual availabilities on 10-50 GHz satellite-earth and aircraft-to-aircraft links, IEE Proc. Microwaves Antennas Propag. 147 (2) (2000) 122–127.

[25] C.L. Wrench, P.G. Davies, J. Ramsden, Global predictions of slant path attenuation on earth-space links at EHF, Int. J. Satell. Commun. Netw. 17 (2–3) (1999) 177–186.

[26] ITU, Attenuation due to cloud and fog, Recommendation ITU-R P.840-5, P Series Radio wave propagation, 2012.

[27] E.R. Westwater, The accuracy of water vapor and cloud liquid determination by dual-frequency ground-based microwave radiometry, Radio Sci. 13 (4) (1978) 677–685.

[28] A.D. Papatsoris, Effect of ice clouds on millimetre-wave aeronautical and satellite communications, Electron. Lett. 33 (21) (1997) 1766–1768.

[29] A. Dissanayake, J. Allnutt, F. Haidara, Cloud attenuation modelling for SHF and EHF applications, Int. J. Satell. Commun. 19 (3) (2001) 335–345.

[30] S.K. Sarkar, I. Ahmad, M.M. Gupta, Statistical morphology of cloud occurrences and cloud attenuation over Hyderabad, India, Indian J. Radio Space Phys. 34 (2005) 119–124.

[31] S.K. Sarkar, A. Kumar, Cloud attenuation and cloud noise temperature over some Indian eastern station for satellite communication, Indian J. Radio Space Phys. 36 (5) (2005) 375–378.

[32] S.K. Sarkar, A. Kumar, Recent studies on cloud and precipitation phenomena for propagation characteristics over India, Indian J. Radio Space Phys. 36 (2007) 502–513.

[33] J.S. Mandeep, S.I.S. Hassan, Cloud attenuation for satellite applications over equatorial climate, IEEE Antennas Wirel. Propag. Lett. 7 (2008) 152–154.

[34] A. Maitra, S. Chakraborty, Cloud liquid water content and cloud attenuation studies with radiosonde data at a tropical location, J. Infrared Millim. Terahertz Waves 30 (4) (2009) 367–373.

[35] V. Mattioli, et al., Analysis and improvements of cloud models for propagation studies, Radio Sci. 44 (2009) 2.

[36] T.V. Omotosho, J.S. Mandeep, M. Abdullah, Cloud attenuation studies of the six major climatic zones of Africa for Ka and V satellite system design, Ann. Geophys. 56 (5) (2014) 0568.

[37] B.K. Mandal, D. Bhattacharyya, S. Kang, Attenuation of signal at a tropical location with radiosonde data due to cloud, Int. J. Smart Home 8 (1) (2014) 15–22.

[38] A.A.R. Kokab, H.A. Edreis, Attenuation (fading) due to clouds South Kordofan (Sudan), IOSR J. Electron. Commun. Eng. 11 (3) (2016) 99–100.

[39] M.R. Sujimol, R. Acharya, K. Shahana, Prediction and estimation of rain attenuation of Ka-band signals, in: 2019 URSI Asia-Pacific Radio Science Conference (AP-RASC), March, IEEE, 2019, pp. 1–4.

[40] A. De, A. Maitra, Radiometric measurements of Ka-band attenuation during rain events at a tropical location, in: 2019 URSI Asia-Pacific Radio Science Conference (AP-RASC), March, IEEE, 2019, pp. 1–4.

[41] D. Nandi, A. Maitra, The effects of rain on Millimeter wave communication for tropical region, in: 2019 URSI Asia-Pacific Radio Science Conference (AP-RASC), March, IEEE, 2019, pp. 1–3.

[42] ITU, Propagation data and prediction methods required for the design of earth-space telecommunication systems, ITU-R P.618-10, 2009.

[43] K. Ulaganathen, et al., Comparative studies of the rain attenuation predictions for tropical regions, Prog. Electromagn. Res. 18 (2011) 17–30.

[44] M. Tamošiūnaitė, et al., Prediction of electromagnetic waves attenuation due to rain in the Localities of Lithuania, Elektronikair Elektrotechnika 105 (9) (2010) 9–12.

[45] F.A. Semire, et al., Analysis of cumulative distribution function of 2-year rainfall measurements in Ogbomoso, Nigeria, Int. J. Appl. Sci. Eng. 10 (3) (2012) 171–179.

[46] Van de Kamp, M.J.L. Max, Statistical analysis of rain fade slope, IEEE Trans. Antennas Propag. 51 (8) (2003) 1750–1759.

[47] O.O. Obiyemi, T.J. Afullo, T.S. Ibiyemi, Equivalent 1-minute rain rate statistics and seasonal fade estimates in the microwave band for South-Western Nigeria, Int. J. Sci. Eng. Res. 5 (1) (2014) 239–244.

[48] S. Das, A. Maitra, A.K. Shukla, Rain attenuation modeling in the 10-100 GHz frequency using drop size distributions for different climatic zones in tropical India, Prog. Electromagn. Res. 25 (2010) 211–224.

[49] K. Ulaganathen, et al., Monthly and diurnal variability of rain rate and rain attenuation during the monsoon period in Malaysia, Radioengineering 23 (2014) 2.

[50] N. Sheng, et al., Study of parabolic equation method for millimeter-wave attenuation in complex meteorological environments, Prog. Electromagn. Res. 48 (2016) 173–181.

[51] H.V. Le, et al., Localized rain effects observed in Tokyo Tech millimeter-wave model network, in: Proceedings of 2013 URSI International Symposium on Electromagnetic Theory (EMTS), IEEE, 2013.

[52] J.S. Mandeep, Comparison of rain rate models for equatorial climate in South East Asia, Geofizika 28 (2) (2011) 265–274.

[53] W. Åsen, C.J. Gibbins, A comparison of rain attenuation and drop size distributions measured in Chilbolton and Singapore, Radio Sci. 37 (3) (2002).

[54] G. Timms, V. Kvicera, M. Grabner, 60 GHz band propagation experiments on terrestrial paths in Sydney and Praha, Radioengineering Prague 14 (4) (2005) 27.

[55] V. Kvicera, M. Grabner, Rain attenuation on terrestrial wireless links in the mm frequency bands, in: Advanced Microwave and Millimeter Wave Technologies Semiconductor Devices Circuits and Systems, InTech, 2010.

[56] C. Capsoni, et al., A new prediction model of rain attenuation that separately accounts for stratiform and convective rain, IEEE Trans. Antennas Propag. 57 (1) (2009) 196–204.

[57] P. Garcia-del-Pino, J.M. Riera, A. Benarroch, Tropospheric scintillation with concurrent rain attenuation at 50 GHz in Madrid, IEEE Trans. Antennas Propag. 60 (3) (2012) 1578–1583.

[58] M.O. Odedina, T.J. Afullo, Analytical modeling of rain attenuation and its application to terrestrial LOS links, in: Southern Africa Telecommunication Networks and Application Conference (SATNAC), 2009.

[59] ITU, Propagation data and prediction methods required for the design of Earth-space telecommunication systems, ITU-R P.618-10, 2009.

[60] J. Vivekanandan, et al., Cloud microphysics retrieval using S-band dual-polarization radar measurements, Bull. Am. Meteorol. Soc. 80 (3) (1999) 381–388.

[61] C. Luo, et al., Channel state information prediction for 5G wireless communications: a deep learning approach, IEEE Trans. Netw. Sci. Eng. 7 (2018) 227–236.

[62] A. Gharanjik, et al., Learning-based rainfall estimation via communication satellite links, in: 2018 IEEE Statistical Signal Processing Workshop (SSP), IEEE, 2018.

The role of IoT in smart cities: Challenges of air quality mass sensor technology for sustainable solutions

13

Alok Pradhan[a] and Bhuvan Unhelkar[b]

[a]*Macquarie University, Sydney, NSW, Australia,* [b]*University of South Florida, Sarasota, FL, United States*

Chapter outline

1 Introduction

Smart and sustainable cities are an imperative for the world's growing population. In their population growth report, the United Nations Department of Economic and Social Affairs [1] predict that the world's population would reach 10 billion by

285

Security and privacy issues in IoT devices and sensor networks. https://doi.org/10.1016/B978-0-12-821255-4.00013-4

2050, with 68% of people living in urban environments, compared with 55% in 2018. This projected increase in population will place higher demands on already strained city resources. Consider, for example, the fact that 60% of the overall energy is being consumed by cities [1]. Additionally the UN estimates that 70% of greenhouse gas and waste is produced by cities as demonstrated [2]. Smart cities aim to reduce such massive energy consumptions and waste production without compromising the living standards of its population. A key element of achieving the aforementioned balance is through active monitoring of the air quality resulting in actionable insights.

Brownfield smart cities[a] in particular have the challenge of managing air quality effectively, as the urban air environment is impacted by traffic networks and older existing industrial technology.

Poor air quality increases the likelihood of respiratory and bacterial diseases for its residents. Particulate matter (PM) in urban air has been widely studied and linked to respiratory issues like asthma, lung diseases, and cardiovascular diseases, especially in children, as demonstrated [3]. Ultrafine particles, less than 0.1 μm in diameter, can penetrate lung tissue and enter the circulatory system, impacting brain development and increasing likelihood of infectious diseases. According to the World Health Organization [4], 4.2 million people die a year around the world from exposure to bad ambient air quality, with almost 1 million deaths being children under 5 years old [5].

Common gas pollutants in urban environments include oxides of nitrogen (NO_x), carbon monoxide (CO), ozone (O_3), and sulfur dioxide (SO_2). Ozone is a secondary pollutant formed by the reaction between atmospheric volatile organic compounds (VOCs) and NO_2 in the presence of ultraviolet radiation from the sun, while the other common gases are predominantly emitted through road traffic and industrial processes. These gases have impacts on the respiratory system, with ozone widely studied to be an aggravator of lung diseases and decreasing lung function [4]. Carbon monoxide can cause cardiovascular and respiratory morbidity in higher concentrations and negative impacts on the central nervous system, particularly in children and the elderly.

The development of the Internet of Things (IoT) in smart cities has been identified as a promising solution to air quality management. IoT is a key element that drives a smart city through interconnectivity and is a consistent term used in literature when defining a smart city. Comprehensive IoT-based urban air quality monitoring using microsensor units (MSU) and low-cost sensors (LCS) can provide precise spatial and temporal air quality data. This form of urban air quality monitoring is being applied in a number of cities around the world. The air quality data collected have provided insights into specific pollutant concentrations in localized areas with high temporal resolution (i.e., data collected every 1 s). These data have assisted in developing mitigation strategies for localized areas.

---

[a]Defined as a city that has IoT and smart technology installed after the city was already established as opposed to greenfield smart cities that are designed and built with connectivity and smart technology.

Applications beyond this now have to be considered to ensure clean, sustainable air quality for citizens. These air quality management applications must be developed in conjunction with emerging technologies to create a holistic management approach. This management approach would consider all influences on urban air quality, such as the urban heat island effect, local traffic, industrial areas, and weather patterns. Data from each of these factors can then be integrated with other urban smart data, including traffic, health, social media, and energy. Applications to improve urban air quality can then be determined using emerging technology, including predictive analytics, machine learning, and digital twin city technology.

This application can be categorized into three tiers—practical, strategic, and regulatory applications. The usability of each tier is determined by the quality of monitoring sensors and data. Practical applications include providing a medium for citizens to monitor air quality in real time and providing alternate routes to pedestrians and cyclists. This type of application is currently being practiced in smart cities. Strategic applications involve installing green walls, integrating air quality data with automated traffic management, and integrating air quality data with urban design planning. This is an area currently under research and has potential to improve air quality in localized areas. Regulatory applications are a substitute to the current methodology of monitoring and reporting air quality against regulated limits. This will include automated comparisons between air quality and regulations, including automated reports and auditing in real time. Furthermore, digital twin cities and artificial intelligence would provide automated mitigation recommendations to ensure regulations are met or provide suggestions of safer pollutant concentration limits depending on weather and other external influences.

2 Background of air quality monitoring sensors in urban environments

This section provides a background on current methods and sensor technology for air quality monitoring. As stated earlier, air quality monitoring is in a state of transition from using expensive and bulky monitoring stations dispersed sparingly to using cheap microsensors that can be deployed densely.

The parameters commonly monitored at ambient monitoring stations are particulate matter (usually PM_{10} and $PM_{2.5}$) and gases NO_x, NO_2, NO, O_3, SO_2, and CO. These parameters are identified as the most abundant pollutants in urban environments that have the most adverse impacts to the health of the general population, especially at moderate to high concentrations [6,7].

Concentrations for particulates and gases can vary within a few meters in an urban environment, particularly with moderate to dense clusters of buildings and traffic movements. Clusters of buildings form canyoning effects where air masses can be trapped and dispersed depending on the positioning of buildings. This can alter air movement and localized air pressure. Building and infrastructure surfaces can also modify air temperatures, all leading to varying particulate matter and gas concentrations over short distances. The flow of traffic can also transport particulates in air masses, depending on traffic volumes, speed, heavy vehicle activity, and tyre-road surface interactions.

Regulation-based ambient monitoring programs have stations installed sparsely across a city, due to capital and maintenance costs. For example, Beijing has approximately 75 regulatory-based ambient monitors spread over 15,000 km^2 [8]—around one station for every 200 km^2. In Los Angeles, there are 26 stations spread over the South Coast Air Basin, of which 20 stations are active [9]. This calculates to one regulatory monitoring station every 850 km^2. As a result of low spatial coverage, regulatory ambient monitoring is conducted only by certified air quality specialists in the areas not represented by regulated ambient monitoring stations. Additionally, Gaussian predictive modeling and computational fluid dynamic models are used to determine pollutant concentrations in areas between monitoring stations. These methods of monitoring and modeling also come at a high cost, and modeling results come with varying uncertainties [8].

MSU and LCS are now being produced and deployed for air quality monitoring in smart and sustainable cities. Due to these sensors being cheap and small, they are being deployed densely in trial monitoring programs. In addition, crowdsourcing air quality programs are also being trialed to gather big data for even larger spatial coverage. The IoT technology in a smart city is then taken advantage of, where the high volume of air quality data is transferred, processed, stored, and communicated in close to real time. However, miniaturizing standardized monitoring methods into MSU and LCS monitoring networks requires known technology and processes to be manipulated and reviewed for precision and validity. As sensors get smaller, either the monitoring methods that have provided precise measurements in traditional ambient monitoring must shrink without compromising accuracy, or new methods must be developed and perfected. In addition, crowdsourced data also carry concern over data validity and accuracy.

Therefore air quality data derived from MSU and LCS sensors and crowdsourcing are currently only used to provide real-time AQI values and indicative information, but not yet for regulatory or research purposes.

The following subsections in this chapter discuss current air quality sensor technology and applications:

Section 2.1: Regulation-based monitoring and sensor technology
Section 2.2: IoT-based monitoring and sensor technology
Section 2.3: Evaluation of regulation and IoT sensors
Section 2.4: Application of urban air quality data

2.1 Regulation-based air quality monitoring

Air quality monitoring across an urban area involves high precision, expensive monitoring equipment installed as stationary stations, with standardized meteorological stations also installed adjacent for reference. These static ambient monitoring stations collect air quality data that are either manually downloaded on-site or automatically uploaded to a central database for reviewing and processing by qualified air quality experts. These air quality stations follow regulatory standards and operate with the purpose of monitoring pollutants against legislative criteria and for research.

Deriving concentrations for each air pollutant requires a specific method involving either chemical, electrical, or photonic technology or a combination of both. The monitoring methods and technology currently used for regulatory monitoring have been developed and continuously improved since the 1980s. International regulatory bodies such as ISO and WHO have consequently produced standardized methodology, which is then adopted by governing bodies into their environmental legislation.

2.1.1 Ambient particulate monitoring

The most accurate and commonly adopted regulatory method for monitoring ambient particulate matter is the gravimetric method. This method involves purpose made filters being preweighed, placed in high-volume air samplers for up to 10 days where air is drawn through the filter at a specific flow rate. The filter is then postweighed, and the concentration is calculated through the measured and known variables. This is a common regulatory method used worldwide to determine ambient air quality and can be used to monitor total PM and PM_{10} and $PM_{2.5}$. The disadvantage of this process is that it is labor intensive as it involves continuous calibration and maintenance of high-volume air samplers and laboratory certified microbalances. Additionally, concentration results are not determined in real time, and actionable air quality management can only be applied after reviewing several results over a period of time.

Continuous monitoring of particulate matter in ambient air for legislative purpose is commonly measured through a tapered element oscillation microbalance (TEOM), which measures particles by using a small vibrating glass tube that oscillates differently as more particles are deposited on a filter that is at the tip of the tube. This oscillation frequency is measured electronically, and a concentration is determined based on this changing frequency, which is proportional to the particulate weight.

A second regulatory-accepted method for the continuous monitoring of particles is a beta attenuation monitor (BAM). A BAM uses the principle that the amount of beta radiation absorbed by a solid particle is relative to its weight. In a BAM monitor, particles are collected along a ribbon, and beta radiation is emitted on precise spot before the particle and after the particle along the ribbon. The difference in intensity in beta radiation in the two locations is then measured, and a mass concentration is derived.

The primary issue faced by both the BAM and the TEOM is varying ambient humidity, as moisture droplets can be mistakenly measured as solid particulates in both methods [10]. As a result the TEOM and the BAM are usually heated at the point of measurement. However, this heating causes a loss in semivolatile particulate matter, resulting in an understated concentration. This is particularly true for continuous monitoring stations near roads or in densely populated areas, where traffic is the main source of air pollutants, as the primary source of atmospheric volatile particulate is road vehicles.

2.1.2 Ambient gas monitoring

Ambient gases are measured through different processes for regulatory monitoring, depending on the gas to be measured. CO_2 is measured through nondispersive infrared (NDIR) technology where an IR light passes IR waves through a tube of sample air toward an optical filter, which is lined up in front of an IR detector. As CO_2 has a unique IR band absorption spectrum, the IR detector can measure the unabsorbed IR waves passing through the tube. The difference between the IR light emitted at one end of the tube and measured by the detector at the other end is calculated, and this difference is directly proportional to the concentration of CO_2 in the tube. Similarly, ozone is measured using an ultraviolet (UV) photometric analyzer, which measures the difference in UV light from one end of a tube to the other, with the tube containing an air sample.

NO_x is commonly measured using chemiluminescence, which uses the natural chemical reaction between NO in the sample air and ozone. The reaction of these elements results in the emission of photons, and the number of photons emitted is directly proportional to the concentration of NO. NO_2 is then calculated based on this reaction and the emission of photons.

These regulatory methods for atmospheric monitoring require regular calibration and maintenance schedules to ensure accurate and valuable results that are collected, which can be then compared with air quality legislative criterion. These methods are also reliable as a result of known uncertainties and ongoing research and development into each technology since the 1980s. This includes the optimal size of the monitoring equipment, including the monitoring chambers, filters, and detectors [11].

2.2 IoT-based air quality monitoring

The technology available to collect, transfer, and store data has generally been well developed for practical use and is in a constant process of improvement through the development of IoT and machine learning technology. But it is the front end (data collection from specific sensors) and back end (the processing and the application of collected and processed air data) of air quality management in smart cities that are in early stages of development and require further attention.

Two primary methods of air quality data collection have been identified in the context of a smart city. These are the following:

a. MSU and LCS sensor deployment by air quality specialists (Section 2.2.1),
b. crowdsourcing and citizen science (Section 2.2.2).

2.2.1 MSU and LCS monitoring

The method that is being commonly trialed in MSU and LCS monitoring for particulates is optical particle counting. This method involves an air stream being introduced into a chamber, where a light or a laser beam is passed through from one end of the chamber to the other, toward an optical detector. The particle scatters or obscures the light or laser beam, and the detector can "count" or measure the energy of scattered light and compare this with the energy of the light initially emitted. The way in which the light is scattered or obscured is directly relative to the size of the particle. Thus the MSU may measure concentrations of total PM, PM_{10}, $PM_{2.5}$, and $PM_{0.1}$.

Gas monitoring methods and sensor technology in MSU and LCS involve the measurement of electric currents or voltage produced by chemical reactions between the sample air and the sensor. For example, miniaturized electrochemical sensors are being trialed across air quality monitoring programs for gas monitoring in many cities, including London, Los Angeles, Zurich, and Beijing [11]. Private air monitoring-based companies have worked closely with research institutes to develop and trial small sensors that can be densely deployed in urban environments. Importantly the goal in trials has been to replicate and exceed the precision and quality of data that is produced by regulatory-based equipment. The methodology of electrochemical sensors is well understood and has been since the 1950s [11]. The challenge today is implementing electrochemical sensor technology in small, cheap, and accurate monitoring devices.

Electrochemical gas sensors measure the concentration of gases based on electric currents produced by oxidation and reduction reactions. An electrochemical sensor typically includes a working electrode, a counter electrode, and a reference electrode. An electrolyte, usually sulfuric acid, is also present in the sensor unit. Air sample passes through a hydrophobic material to remove moisture and then reaches the working electrode. The reaction between the sample air and the working electrode creates an oxidation or reduction reaction, creating a current between the working electrode and the counter electrode. The size of this current is measured and is proportional to the concentration of the target gas.

Another method that is currently being researched and tested as a mass sensor approach to gas monitoring in smart cities is using electrical responses from elements, particularly measuring the variations in conductivity of metal oxide [12]. Metal-oxide semiconductor (MOS), electrochemical, and polymer technology are also being used in trials of microsensors in air quality monitoring. Ground level ozone and NO_2 in particular are being monitored through trials with MSUs using MOS technology. This method was discovered in the 1970s, but due to sensors being sensitive to interference gases, as well as having issues in providing precise NO_2 and ozone concentrations outside a small ambient temperature and humidity range, the technology was not used extensively in air or emissions monitoring.

MOS uses a heated metal oxide using a small resistive heater, which causes oxygen to bond to it at the boundary between the surface and bulk layer. This draws electrons away from the semiconductor surface, changing the sensor resistance. The metal oxide resistance varies specifically to changes in the composition of elements in the air. From this change in resistance, a concentration for NO_2 can be derived at a small-time resolution. MOS is a relatively new method to monitor air quality, as the technology is still being improved to counter high sensitivity to temperature and humidity.

Due to the sensitivity of the metal oxide to the composition of the air, calibration is required frequently and only in the monitoring location. This method also makes it imperative to apply humidity corrections to the data as moisture can impact the semiconductor surface interaction with the ambient air.

Yet another technology currently in research for MSU and LCS gas monitoring includes the use of graphene, transitional metals, and the use of organic materials. These materials have unique surface interactions with elements in the air, especially at very low concentrations, and therefore are being researched further. Cost, operating temperatures, and reliability are the primary challenges identified through feasibility studies and laboratory tests [13]; however, heterogeneous structures that combine materials mat assist with these challenges. In addition, these materials can detect trace toxic gases such as hydrogen sulfide (H_2S) and sulfur dioxide and the common gases in urban environments, which would be ideal for monitoring in the industrial areas of a city.

2.2.2 Crowdsourcing and citizen science

Crowdsourcing applications in air quality monitoring and management are even a younger field of research than monitoring using MSU and low-cost sensors, with limited publications and research in the use of passive crowdsourcing and formal citizen science projects [14]. Citizen science-based projects use MSU developed by air quality monitoring equipment manufacturers, by providing volunteers with the sensors, monitors, coding programs, and instructions for the operation and maintenance of the MSU. Modern air quality monitoring leaders such as Alphasense and Aeroqual are trialing such programs to determine the effectiveness of using crowdsourcing by selling cheap units to volunteers, collecting data from as many sources as possible, and providing useful air quality statistics back to the public, including AQIs for each pollutant monitored. A goal of these assessments is also to increase awareness and interest and educate the importance of air quality in urban environments.

2.3 Evaluation of regulation- and IoT-based air quality sensor technology

There are seven parameters that have been identified in literature, which can be applied to assess the effectiveness of air quality sensors [13,15,16]. These seven parameters are listed and described in the succeeding text:

1. *Sensitivity* refers to the minimum concentration detection limit of a sensor, which is crucial in the measurement of trace elements in ambient air.
2. *Selectivity* refers to the ability of a sensor to detect specific elements within an air sample and discount others.
3. *Response time* refers to the time it takes from collecting an air sample to deriving a concentration. In the case of IoT in smart cities, this is a key element in air quality management. Response time is also important for a network of air quality sensors as it has direct implications on communication bandwidths.

Table 1 Monitoring technique evaluation against seven performance variables.

Monitoring technique	Sensitivity	Selectivity	Response time	Reversibility	Volume of data	Power consumption	Cost
Regulation based ambient stations	4	5	1 (particulates) 4 (gases)	4	2	1	1
Micro sensor units	3	3	5	4	5	4	5

4. *Reversibility* refers to whether the parameters that are collected and measured through the sensor can return to their same form as before they were measured.
5. *The volume of data* able to be collected.
6. *Power consumption* impacts the value and performance of the sensors and the monitoring network.
7. *Cost* of the sensors, maintenance, and the available IoT infrastructure.

The challenge of developing small and cheap sensor technology is to reproduce and improve on the methods of accurate and precise measurement techniques that have been researched and applied in regulation-based monitoring stations while miniaturizing the technology. Regulation-based monitoring sensors perform well in selectivity, sensitivity, reversibility, and response time. However, volume of data and power consumption and costs present challenges with regulation-based sensors.

Table 1 presents an evaluation of reference-based and IoT-based monitoring against the seven performance variables for air quality sensors, with a rating out of five. Green represents good performance (4–5), orange represents average performance (2–3), and red represents poor performance (0–1).

2.4 Applications of urban air quality data

Data from regulation-based monitoring are predominantly used for legislative compliance assessments and research in health and environmental implications. Regulation monitoring stations also provide almost real-time AQIs for each pollutant measured. The AQI is an indicative value, providing information to the general public on whether it is safe to spend extended time outdoors and undertake vigorous exercise or if there are any hazards for those already with respiratory issues. Each regulatory body develops their own AQI scale and presentation, but it is commonly determined on the relative concentration of each gas or particulate parameter against various health criteria [17]. The AQI consists of a scale of good to dangerous air quality for human health and is developed and quality checked by qualified specialists.

The application of air quality big data from MSU and LCS in a smart city context needs to be addressed next. Currently, data from IoT-based sensors are also used for AQI communication with the public through IoT infrastructure. The advantage of AQI data produced from MSU- and LCS-based monitors over regulation-based monitors is the spatial coverage, as discussed at the start of this section. However, a disadvantage can be the validity of the data collected, as data verification can become an issue with large amounts of heterogeneous data.

The deployment and management of MSU networks can be three to five times less expensive than regulation-based monitoring station networks [16], with the advantage of collecting mass data with a high time resolution and uploading this data for immediate processing to a centralized processing cloud database. The development and availability of wireless local area network technology, the marketing of low-cost access points, wireless network adapters, and wireless bridges create a cheap and readily available option for connectivity and data transfer. This creates air quality networks to be dense and highly connected providing real-time communicable information to stakeholders.

In current trials of MSUs and LCSs for air quality monitoring, monitors are being deployed on light posts along streets, on traffic lights, building corners, trash cans, and bus stops. MSU networks are also being deployed on government buses, trams, local trains, and cycles in bike-share programs. Dynamic monitoring has the advantage of providing real-time data continuously throughout a city. For example, a monitor traveling on a bus will collect and send data to a centralized database continuously, with time resolutions as low as 1 s. These data will include the monitor's spatial coordinates at any given time and the gas or particulate concentrations at that precise location [18].

The networking of air quality sensors is formed by their ability to transfer collected data immediately to a centralized cloud or fog platform. The transfer of this data is commonly carried out by low-cost Wi-Fi chips, such as ESP8266 or RN1723-1 microchips, and open-source hardware and software ecosystems using mesh networking [19]. Data are transferred to a fog or cloud platform for processing through these devices. Once air quality-based algorithm and data management has occurred, the data are then distributed to programs or apps to the public.

Fig. 1 presents the process chart of air quality management within a smart and sustainable city, as presented in this chapter. Future applications of IoT-based air quality monitoring are discussed in a later section within this chapter.

3 Challenges of air quality monitoring and management in urban environments

Challenges to air quality monitoring, especially in urban environments, can be listed as follows:

- lack of spatial variability of standardized ambient air quality monitoring stations;
- miniaturization of air quality sensors resulting in data accuracy and validity of MSU and LCS technology due to environmental factors;
- data validity, biases, and security concerns of crowdsourced air quality monitoring data;
- lack of transparency of MSU and LCS sensor methodology due to patency of technology and methods.

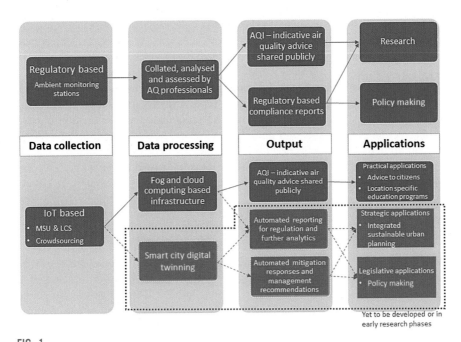

FIG. 1

Smart city air quality data management and application.

3.1 Lack of spatial variability

Dynamic changes in particulate and gas concentrations (specifically that of PM_{10} and $PM_{2.5}$) over short distances are not accurately captured by traditional air quality monitoring stations. While MSUs and LCSs can address this lack of coverage through dense monitoring networks, they pose a challenge in accuracy and data validity in comparison with traditional monitoring methods.

3.2 Environmental challenges

Accuracy issues in MSU and LCS are partly formed due to misreadings of concentrations in high humidity and in extreme temperatures, as well as from cross-interference from other ambient gases [20]. For example, a review on currently operating MSU that contributes to providing an AQI determined that particulate sensors developed by private firms Sharp, Shinyei, Nova, and Dylos were found to provide precise monitoring results only if site-specific field calibrations were performed regularly (weekly to monthly) and only worked more effectively in higher polluted areas. Detection of particulate matter at lower concentrations was not achieved by most sensors. Humidity and temperature variations were found to influence the sensors, particularly when there is a low concentration of particulates in the air [16].

To overcome the abovementioned challenges, developers create sensors and particularly optical counting microsensors with a humidity correction component. Uncertainty factors are commonly implemented into algorithms to correct pollutant concentrations for uncontrolled variables but need vigorous analysis when used by sensors for air quality in a smart city context [21]. Uncertainty factors are dependent on meteorological conditions in the specific monitoring location for each sensor and the methodology used to measure ambient air particulate and gas concentrations.

Correction components and uncertainty factors, however, require regular on-site calibrations to maintain precise measurements [21]. On-site, manual maintenance and calibration of a mass number of sensors spread throughout a city then become an expensive exercise and defeat the purpose of automated, IoT-based monitoring.

3.3 Crowdsourcing and citizen science challenges

As a new field of research, crowdsourcing data for air quality monitoring provides ongoing challenges. Data validity forms the biggest challenge in crowdsourced air quality data. Volunteers may not have a strong enough knowledge base or simply lose interest in maintaining the quality of the sensors and monitors and therefore cease to conduct regular calibrations and maintenance. As stated earlier, calibrations are crucial to ensure valid data to count for the specific location of the sensor, as well as humidity and temperature corrections. To account for this, software updates and calibrations can be performed remotely through IoT infrastructure similar to updates rolled out to mobile phones and computers; however, this can be void if sensors are not maintained properly or located in appropriate locations.

Volunteers may not have appropriate outdoor areas for monitors to be installed and therefore can be affected by local atmospheric effects such as eddy currents, indoor-outdoor air quality interactions, nearby building ventilations, and ground effects, where the sensors are not far enough aboveground to counter abnormal particle imbalances.

Crowdsourcing raises more challenges for data validity, where biases and misinformation need to be carefully filtered to get valuable results. A further step to this would be clarifying context and biases from different social media and outlets. For example, photo-based social media like Instagram contain heavily filtered and machine-aided photographs that can alter the context making it difficult to interpret the image [14].

Biases can also form with the population of interest or volunteers having similar socioeconomic status, motivations, and education. Additionally, very active users of social media and blogs who would be interested in environmental management and air quality tend to be younger professionals [22].

Access to mobile phones and Internet can also prevent certain groups of society to be able to participate in crowdsourcing.

Security issues and ethics are also key challenges in crowdsourcing. Mining data without users' consent can be a violation and can increase online privacy concerns [23]. Data-mining social media, blogs, websites, etc. for air quality-related content are overall less intrusive than survey-based studies, as direct answers from questions are not being published, but rather key words, ideas, and air quality data from publicly available information [22]. The handling of the ownership of data and resulting profits of data for research can also be a key challenge and would need to be clear from the outset with volunteers in citizen science-based data collection.

3.4 Lack of transparency

Another challenge currently facing mass sensor network technology is the transparency in technical methods being trialed to monitor air quality data. There are currently many trials of mass air quality sensor networks being tested in cities around the world. However, the technology, calibration techniques, and algorithms for data processing are largely unknown due to sensors being developed by private firms who have patented their processes as they go through research and development phases [24].

Part of the reason for little to no transparency in specific methods is due to monitoring processes not yet being standardized. As MSUs and low-cost sensors are currently only used to communicate indicative AQIs to the general public, there is yet a need to regulate the monitoring methodology. The onus is therefore on the sensor manufacturer to ensure highly accurate results are being communicated. Additionally, passive sensors are mostly being used for air quality monitoring and connectivity, such as Arduino and Raspberry Pi-type open-source hardware. While these are cheap and robust, they do not provide the high technical standards required for precise regulation-based air quality monitoring.

To address data quality, the US Environmental Protection Agency (USEPA) is in the process of developing standardized methodology for MSU and LCS technology. This has involved assessing and auditing technology and methodologies against the seven performance variables outlined earlier in laboratory and field conditions. This assessment is for the whole air monitoring cycle from the sensor type, monitoring technology, data transfer, data processing, and data storage and communication. The assessments have so far concluded that the current systems and technologies are fit for specific purposes of providing indicative AQI, but more work is required to enhance data quality for compliance testing and air quality management [16].

3.4.1 IoT-based air quality monitoring trial programs

Each of the monitoring methods of particulates and gases detailed in Sections 3.1–3.4 is being tested with different parameters and materials, in various trials, across different cities. From 2012 to 2019, several field and laboratory tests have been carried out on various particulate and gas sensor technology.

Successful sensors in trials have been the Alphasense models OPC-N2 and OPC-N3, which have provided a good correlation with regulation instruments run concurrently during the trial period, particularly for $PM_{2.5}$ and PM_{10}. The specific technology in this optical counting sensor compared with other optical sensors is difficult to evaluate, due to patents. Alphasense sensors also performed well using gas sensors, as did Aeroqual and ClairClip sensors [24].

Most, if not all, trials of IoT-based air quality monitoring are being undertaken throughout Europe, North America, and Asia. For example, in London, the Breathe London initiative involved a trial of stationary mass sensors across the 32 boroughs of London, from late 2018, and will run until late 2019. The program deployed approximately 100 Alphasense AQMesh pods across London, which contained NO_x, PM_{10}, and $PM_{2.5}$ sensors. These pods were installed on light posts, taking 10-s measurements on a 1–15-min time average. The pods contained NDIR sensor for CO_2, optical scattering for particulates, and electrochemical-based methods for other gases. Early assessments of the AQMesh so far have revealed that there are influences from varying humidity and temperatures that may have had some impacts on the results [20].

A trial in the Los Angeles basin by Aeroqual [25] was installed in 2017, in response to the mandate implemented by the California state government, to have community-level air quality monitoring installed in the area by mid-2019. This project was deemed technically and economically unfeasible; however, Aeroqual did install several LCS in the area to monitor O_3, NO_2, and $PM_{2.5}$. This project has provided valuable air quality data for the community; however, the management of mass LCS over a large area proved difficult without a standardized process.

4 Applications, initiatives, and future direction

Potential applications of smart technology in air quality IoT frameworks are discussed in the succeeding text, as follows:

Section 4.1: Integration of sensor data and passive crowdsourced data.
Section 4.2: Air quality data application framework.
Section 4.3: Smart city digital twinning.

4.1 Integration of sensor data and passive crowdsourced data

Passive crowdsourcing is a form of data collection that involves accessing social media outlets, blogs, websites, and other publicly available data to gather any information with regard to air quality. The idea of extracting information from social media is not a new concept, as marketing firms, advertisers, and emergency departments have been taking advantage of this as a data resource for over a decade. However, the concept of using social media for environmental management is relatively new and has recently been gaining momentum in academic literature [14].

Passive crowdsourcing does raise security concerns and data validity issues, as discussed in the previous section. However, unlike sensor data, people's interactions with nature, emotional affinity, and responsiveness to air quality issues can be collated and analyzed against collected air quality data from MSU and LCS. This integrated information can be highly informative in smart and sustainable cities and addresses the aim to improve the standard of living for its residents, of which health and happiness are key determinants.

The integration of air quality sensor and passive crowdsourced data can be taken advantage of, to analyze the following:

1. Behavior and attitudes toward mass air pollution events or noticeable bad air quality (e.g., from bushfires, mass traffic jams, high-wind events, low air quality air masses trapped under inversion layers).
2. "Social sensing" analysis, similar to "remote sensing," for example, areas from which social media posts are negative toward air quality on a more frequent basis, could lead to more advanced monitoring or air quality management such as traffic diversions, "no go zones" for pedestrians or cyclists, air flow management to remove pollutants, and health implications in certain zones. Attitudes and behavior from varying demographics can also be analyzed.
3. Behavior and responsiveness differences during clean air versus bad air quality events could contribute toward the impact of air quality on the contentment of people within urban environments.

Data collected from passive crowdsourcing combined with data from MSU and LCS monitoring networks can be taken advantage of by transmitting both streams of data to a fog processing database, prior to cloud processing and storage. Fog computing creates a more efficient process in this ecosystem, as it consists of a decentralized processing and storage system, which assists with processing large volumes of data. Large volumes of social media and air quality data can be processed and then stored quickly via localized cloudlets, providing much faster real-time updates. Each node that forms part of the fog system could represent a local area of the city, with air quality data and passive crowdsourced data combined in those specific nodes, enabling an efficient and well-organized processing infrastructure before being transmitted to a cloud for further processing and distribution as required.

Fog computing has the added advantage of adding extra ICT security, as the data collected is processed by a complex system that consists of a distribution of nodes (further discussion by Javidi et al. [26]). Node distribution in fog computing is a particularly important component as a system like this would be compiling important location-based data.

The combination of heterogeneous data from social media and data from sensors can take advantage of the progress of machine learning and artificial intelligence. Machine learning-based algorithms could compare the behavioral response of residents with the quantitative air quality data collected and provide detailed analytics. Accordingly, this information would then be supplied to decision makers to implement policy to improve air quality management and the standard of living for residents.

The role of deep computational learning will no doubt also plays a part in analyzing passive crowdsourced data and air quality data analytics, and along with standardized air quality monitoring methods using MSU and LCS and crowdsourcing, air quality in urban areas can be managed effectively through policy and strategic applications.

4.2 Air quality data application framework

Air quality monitoring using MSU and LCS has created the opportunity for valuable IoT applications, involving mass amounts of air quality data to be collected and analyzed. Information gathered can then be applied by informing citizens of specific areas of good or bad air quality, informing hospitals of potential air quality health-related events, informing engineers of appropriate air ventilation designs, and informing decision makers on traffic, services, and environmental policies.

Historical and future trends of the air quality in specific council areas, suburbs or streets, can be processed through big data analytics, and solutions to improve air quality in these areas need to be developed. Data collected can move from being used as indicative air quality indices for a city's residents to also being used by decision and policy makers. That is, with monitoring technology improving and big data availability, actions can be implemented by air quality experts and decision makers to improve air quality.

As stated, ambient air quality monitoring in cities is in a transitional state of moving from being carried out solely for regulatory and research purposes to also being used for communication with city residents.

The potential applications of air quality big data need to be developed simultaneously with the development of LCS technology and methods for data mining. The purpose of monitoring air quality should be to ensure that the composition of the urban air is not in any way harmful to people within the city or the natural environment and to otherwise improve air quality. To achieve this a three-tiered air quality data application hierarchy is promoted, as outlined in Table 2.

4.2.1 Practical applications

The first tier is implementing practical application. This application of data is currently practiced and involves actions such as providing real-time, indicative air quality data publicly through websites and apps. This includes informing people of any hazardous air quality warnings and recommendations to those people most sensitive in real time, through IoT-based applications.

Practical applications also include providing rerouting information for people within a city, to avoid poor air quality areas. This can be valuable for pedestrians and cyclists who travel close to industrial zones or high traffic roadways. There is a high potential for this type of application to be implemented, especially in cities where local AQI information is already accurately developed [17]. An example of this type of practical application is the OpenSense pilot project in Zurich, where almost real-time concentrations of ultrafine particles were mapped and shared through an app, measured by dynamic LCS [27].

Practical applications inform citizens and residents of real-time air quality in local areas, including information on suggested levels of physical activity, notable hazards, and any precautions to consider. Currently, this is the level at which big air quality data collected from MSU and LCS are currently applied and trialed.

Table 2 Applications of air quality big data.

Application	Description	Example
Practical	Indicative air quality data available publicly	• OpenSense project in Zurich • Array of Things project in Chicago • Real-time AQI and air quality information via smart phones and Internet
Strategic	Application of programs and incentives to improve air quality	• Real-time management of air quality in sensitive areas of a city (schools and hospitals) at peak times via traffic lights • Integration of passive crowdsourced data with sensor data for specific location-based strategies • Sustainable designs implemented in urban planning, specific to results from air quality sensors
Legislative	Regulatory and policy changes to directly improve air quality	• ULEZ in London • Licensing for pollutant emission sources • Standardization of MSU and LCS equipment and methods • No car zones, electric vehicle only zones, pedestrian only zones

4.2.2 Strategic applications

The second tier is strategic applications, which involves using data to implement programs to directly improve air quality. Data collated from LCS and MSU can be analyzed through big data analytics, and the results can be used to develop strategies in local areas or city wide to improve air quality. An example of a strategic application would be the communication of air quality sensors with traffic lights through IoT infrastructure. Traffic lights could guide traffic to move constantly in certain areas where MSUs and LCS have detected poor air quality.

A similar strategic application could be applied around schools, hospitals, and other sensitive areas; ensuring traffic avoids high-density pedestrian locations during peak times.

Other examples of strategic applications of air quality data include identifying areas that can be converted to electric vehicle-only zones, cycle-only zones, green spacing, or pedestrian-only zones. These zones can be implemented through urban planning designs as sustainable development, informed by air quality data. Once developed the integration of passive crowdsourcing with quantitative air quality data from MSU and LCS can be used to determine these strategic applications. These data would assist in determining specific locations within a city where air quality is more of a concern for residents and needs to be addressed.

Strategic applications can also take form as predictive analytic tools. Siemens has developed smart technology using neural networks, which can predict air quality up to 5 days in advance. The machine learning-based technology identifies past local meteorological conditions, transport networks, and air quality conditions and uses neural networks to predict future trends. The technology was trialed in London with 150 MSU deployed across Central London and resulted in 90% accuracy in predictions [28].

4.2.3 Legislative application

The third tier is legislative application. As with strategic applications, regulatory-based changes can only be made with definitive and precise air quality data from standardized monitoring equipment and monitoring methods. Legislative applications include licensing pollutant emitting businesses to limit their emissions, forcing new technology or relocation from highly polluted areas of the city. Mandating no-vehicle zones, rather than tolling vehicles, would be another likely legislative application. Mandating more pollution sinks, air quality treatments, and polluter pay schemes can also be applied using data collected from MSU, LCS, and crowdsourcing.

The implementation of the ultralow emission zone (ULEZ) in Central London is an example of a legislative application, which implements a toll for vehicles entering Central London. While the ULEZ was not implemented via air quality data collected from LCS and MSU, it is the type of application that could be developed and monitored through mass sensor technology and IoT, once air quality data from these sources become more precise.

Current air quality research has focused on the collection of big and the most effective methods to collect accurate and valuable information. Due to the uncertainties surrounding the precision of the monitoring method and therefore air quality data, the application of results only goes so far as indicative information for residents. The next step is to simultaneously apply standards to determine local air quality policy, directly relevant to the spatial distribution of the air quality monitors.

Applications of machine learning and AI have already begun with the development of neural networks to provide accurate predictions of air quality. This will assist decision makers to act accordingly. The pilot project of the Siemens developed air quality predictive technology in London learnt to account for impacts on work days, weekends, event days, sporting events in certain areas, and other events and how air quality patterns change accordingly [28].

In a similar project in Guangzhou City, the City Air Management smart technology also developed by Siemens uses MSU and LCS to create an air quality prediction dashboard, again highly accurate and self-learning. In addition to predicting air quality levels in the future, the software also recommends mitigation measures that can be acted upon at short notice, such as implementing free transportation for a short period of time [29].

Smart technology is a key driver to improving air quality once MSU and LCS technology can provide accurate air quality readings.

4.3 Smart city digital twinning

Digital twinning technology can also be implemented in forecasting air quality and developing preventive applications for poor air quality. Using the huge amount of big data collected from MSU, LCS, and crowdsourcing data, the data can be processed to create a replica of the real world and its components [30]. In terms of digital twinning a smart and sustainable city, components like transport and population growth are already being developed to determine accurate forecasts in demographics and transport strains. However, digital twinning for environmental management is relatively new with limited literature available.

Predicting air quality and applying the most effective strategies to improve air quality through digital twinning will require a large array of data including historical air quality data, building, road and rail infrastructure, industrial sources of pollution, meteorological conditions, energy use, vegetation cover, and other natural features, as well as population growth and traffic patterns. Accurate data from each of these industries are required to determine accurate air quality predictions. While IoT involves open, real-time connectivity between different concepts, digital twinning involves input data and sensor technology, a method to collect and store data for a large idea, like an entire city, which can then be used in an advanced way to test and build ideas in a virtual environment [30].

A key requirement in digital twinning is multidisciplinary collaboration, as various disciplines are required to feed into a model to determine their impact on each other. Using a digital twin city model, traffic buildups in one area can be predicted and mitigated through the model and neural networks. Solutions and development can be tested within the model, and real-time signals can be sent to the physical component of the city for actions to be carried out [31].

Urban life quality is already being tested through a digital twin of the City Antwerp in Belgium. Traffic, noise, and air quality components are being predicted in advance, with recommended options for mitigation or strategic applications presented to policy makers who simply need to push a button to implement the action [32]. This digital twin is continuously being developed (and self-learning), with sensor data still being absorbed by the model and trials being carried out. However, it has so far been deemed successful in terms of urban strategy planning.

The potential for digital twinning being applied for air quality management within a city is enormous; however, the fundamentals first have to be developed to be valid and precise, that is, the validity and precision of air quality sensor technology. Along with the applications of machine learning in creating automated mitigation strategies for air quality management, digital twinning for improving urban air quality is an area requiring further research.

5 Discussion and findings

The primary goal of conducting air quality monitoring in a smart and sustainable city is to improve air quality for its citizens. It is clear there will be an increase in demand for world's city resources, and therefore the development of IoT needs to be taken advantage of to create sustainable, clean air within urban environments.

Research efforts have focused on the use of IoT frameworks to automatically communicate accurate AQI data to residents. Trials and research of different MSU and LCS are ongoing; however, this chapter has shown that the sensors are suitable only to provide indicative air quality measurements that can be communicated in real time through IoT infrastructure.

This chapter has deduced that crowdsourcing is a valuable source of data collection in a smart and sustainable city context. In terms of air quality management, people's interactions with nature, emotional affinity, and responsiveness to air quality issues and events can be collated. In the context of a smart and sustainable city, this is vital as people's standard of living is formed partly by their health and happiness. Air quality management applications could be applied according to big data and predictive analytics performed on collated sensor-based data and passive crowdsourced data. This form of data application can be strategically managed to achieve the overall outcome of improving air quality within urban environments.

There are pilot and trial projects in place in cities across the world linking air quality monitors, with other city infrastructure. However, this chapter has deduced that the accuracy of sensor results needs to improve for this and other strategic applications to be effectively managed. Moreover, air quality sensor technology needs to perform at a higher level, so results can be used for regulation-based monitoring, which will force standardized methods for IoT-based air quality monitoring. This will result in a focus more on improving and maintaining sustainable urban air quality, rather than the current focus on achieving compliance and maintaining equipment.

Beyond this, this chapter has shown that the future direction of air quality management lies in a holistic, digital approach. There is an opportunity for air quality management to be conducted through digital twin cities, which can incorporate all sensor collected big data in a smart and sustainable city. Urban air quality is easily influenced by a variety of factors from infrastructure, weather, traffic, industries, and energy use. It would therefore be ideal to explore the capability of managing air quality through a digital twin, where all urban emission sources and parameters would be considered in one real-time model.

6 Conclusion

Holistic management of air quality in cities is crucial to set a reasonable standard of living for its citizens. There is now a potential for this approach with the use of IoT technology, combined with ML and AI technology in urban environments.

The air quality sensors and monitoring technology now available allow the collection of air quality big data. This includes both qualitative and quantitative data. This chapter provides an analysis of the methods and quality of air quality data and proposes a three-tiered approach to managing urban air quality. While practical applications such as real-time air quality information and warnings are already being applied, strategic and regulatory applications are the next logical steps in making the most of air quality data collected. A potential platform using IoT and machine learning technologies is digital twin smart cities, which can provide analysis and, in the future, mitigation strategies based on urban data. However, the quality of current air quality data inputs into such platforms still needs refining to be accurate. This is due to the transition of air quality monitoring apparatus, from large spatial and temporal scales to almost street-by-street scale.

There is the potential for further research into how IoT-based air quality data can be refined further, to be strategically applied in a way that can reduce air quality impacts in urban environments.

References

[1] United Nations, Department of Economic and Social Affairs, Population Division, World Urbanization Prospects: The 2018 Revision (ST/ESA/SER.A/420), United Nations, New York, 2019.

[2] International Organization for Standardization, ISO and Smart Cities, ISO Central Secretariat, Geneva, Switzerland, 2017.

[3] H.-Y. Liu, D. Dunea, S. Iordache, A. Pohoata, A review of airborne particulate matter effects on young children's respiratory symptoms and diseases, Atmosphere 9 (2018) 150.

[4] World Health Organisation International Agency for Research on Cancer, IARC: Outdoor air pollution a leading environmental cause of cancer deaths [Press Release no. 221], Available at: https://www.iarc.fr/wp-content/uploads/2018/07/pr221_E.pdf, 2013. Accessed 5 September 2019.

[5] J. Lelieveld, A. Haines, A. Pozzer, Age-dependent health risk from ambient air pollution: a modelling and data analysis of childhood mortality in middle-income and low-income countries, Lancet Planet. Health 2 (7) (2018) E292–E300.

[6] J. Fenger, Urban air quality, Atmos. Environ. 33 (29) (1999) 4877–4900.

[7] D. Mage, G. Ozolins, P. Peterson, A. Webster, R. Orthofer, V. Vanderweed, M. Gwynne, Urban air pollution in megacities of the world, Atmos. Environ. 30 (5) (1996) 681–686.

[8] H. Zhao, Y. Zheng, C. Li, Spatiotemporal distribution of PM2.5 and O3 and their interaction during the summer and winter seasons in Beijing, China, Sustainability 10 (12) (2018) 4519.

[9] California Air Resources Board, Quality Assurance Air Monitoring Sites by County, [online]. Available at: https://ww3.arb.ca.gov/qaweb/countyselect.php?c_arb_code=70, 2019. Accessed 14 September 2019.

[10] Z. Bai, J. Han, M. Azzi, Insights into measurements of ambient air $PM_{2.5}$ in China, Trends Environ. Anal. Chem. 13 (2017) 1–9.

[11] J. Engel-Cox, N.T.K. Oanh, A. van Donkelaar, R.V. Martin, E. Zell, Toward the next generation of air quality monitoring: particulate matter, Atmos. Environ. 80 (2013) 584–590.

[12] J. Burgues, S. Marco, Low power operation of temperature-modulated metal oxide semiconductor gas sensors, Sensors 18 (2) (2018) 339.

[13] A.H. Khan, M.V. Rao, Q. Li, Recent advances in electrochemical sensors for detecting toxic gases: NO_2, SO_2, and H_2S, Sensors 19 (4) (2019) 905.

[14] A. Ghermandi, M. Sinclair, Passive crowdsourcing of social media in environmental research: a systematic map, Glob. Environ. Chang. 55 (2019) 36–47.

[15] J. Smalley, F. Vallini, A. El Amili, Y. Fainman, I.N. Da Silva, R.A. Flauzino (Eds.), Photonics for Smart Cities, Smart Cities Technologies, IntechOpen, 2016. https://doi.org/10.5772/64731.

[16] L. Morawska, P.K. Thai, X. Liu, A. Asumadu-Sakyi, G. Ayoko, A. Bartonova, A. Bedini, F. Chai, B. Christensen, M. Dunbabin, J. Gao, G.S.W. Hagler, R. Jayaratne, P. Kumar, A.K.H. Lau, P.K.K. Louie, M. Mazaheri, Z. Ning, N. Motta, B. Mullins, M.M. Rahman, Z. Ristovski, M. Shafiei, D. Tjondronegoro, D. Westerdahl, R. Williams, Applications of low-cost sensing technologies for air quality monitoring and exposure assessment: how far have they gone? Environ. Int. 116 (2018) 286–299.

[17] F. Ramos, S. Trilles, A. Munoz, J. Huerta, Promoting pollution-free routes in smart cities using air quality sensor networks, Sensors 18 (2018) 2507.

[18] S. Kaivonen, E.C.H. Ngai, Real-time air pollution monitoring with sensors on city bus, Digit. Commun. Netw. 6 (2019) 23–30. Available: https://www.sciencedirect.com/science/article/pii/S2352864818302475. Accessed 10 September 2019.

[19] M.B. Marinov, I. Topalov, E. Gieva, G. Nikolov, Air quality monitoring in urban environments, in: 39th International spring seminar on electronics technology (ISSE), Pilsen, Czech Republic, 2016 [online]. Available at: https://ieeexplore.ieee.org/abstract/document/7563237. Accessed 7 September 2019.

[20] P. Wei, Z. Ning, S. Ye, L. Sun, F. Yang, K.C. Wong, D. Westerdahl, P.K.K. Louie, Impact analysis of temperature and humidity conditions on electrochemical sensor response in ambient air quality monitoring, Sensors 18 (2) (2018) 59.

[21] L. Spinelle, M. Gerboles, M.G. Villani, M. Aleixandre, F. Bonavitacola, Field calibration of a cluster of low-cost commercially available sensors for air quality monitoring. Part B: NO, CO and CO_2, Sensors Actuators B Chem. 238 (2017) 706–715.

[22] V. Heikinheimo, E. Di Minin, H. Tenkanen, A. Hausmann, J. Erkkonen, T. Toivonen, User-generated geographic information for visitor monitoring in a National Park: a comparison of social media data and visitor survey, Int. J. Geo-Inf. 6 (3) (2017) 85.

[23] M. Zimmer, K. Kinder-Kurlanda, Internet Research Ethics for the Social Age, Peter Lang, New York, USA, 2017, pp. 177–187.

[24] B. Feenstra, V. Papapostolou, S. Hasheminassab, H. Hang, B. Der Boghossian, D. Cocker, A. Polidori, Performance evaluation of twelve low-cost $PM_{2.5}$ sensors at an ambient air monitoring site, Atmos. Environ. 216 (1) (2019) 116946.

[25] Aeroqual Project Case Study, Los Angeles Field Evaluation Case Study [online], Available at: https://d2pwrbx99jwry6.cloudfront.net/wp-content/uploads/Case-Study-LA-Community-Air-Monitoring-Network.pdf, 2017. Accessed 5 September 2019.

[26] G. Javidi, E. Sheybani, L. Rajabion, Fog computing: a new space between data and the cloud, Cutter Bus. Technol. J. 30 (10) (2017) 54–57.

[27] M. Urech, Visualize volumes of air quality data, ETH University, Department of Information Technology and Electrical Engineering, 2017. [online]. Available at: https://pub.tik.ee.ethz.ch/students/2017-FS/MA-2017-06.pdf. Accessed 15 September 2019.

[28] S. Webel, Smart Cities: Forecasting Software that's a Breath of Fresh Air, Siemens Research and Technologies, 2018. [online]. Available at: https://new.siemens.com/global/en/company/stories/research-technologies/folder-future-living/smart-cities-air-pollution-forecasting-models.html. Accessed 22 September 2019.

[29] A.G. Siemens, Siemens software solution helping cities improve air quality, [online]. Available at: https://press.siemens.com/global/en/pressrelease/siemens-software-solution-helping-cities-improve-air-quality, 2018. Accessed 10 September 2019.

[30] N. Mohammadi, J.E. Taylor, Smart city digital twins. in: 39th International Convention on Information and Communication Technology, Electronics and Microelectronics (MIPRO), Honolulu, Hawaii, 2016. https://doi.org/10.1109/SSCI.2017.8285439.

[31] N. Mohammadi, J.E. Taylor, Devising a game theoretic approach to enable smart city digital twin analytics, in: Proceedings of the 52nd Hawaii International Conference on System Sciences, Honolulu, Hawaii, 2019 [online]. Available at: https://scholarspace.manoa.hawaii.edu/handle/10125/59639. Accessed 13 October 2019.

[32] IMEC, Imec and TNO launch Digital Twin of the city of Antwerp, [online]. Available at: https://www.imec-int.com/en/articles/imec-and-tno-launch-digital-twin-of-the-city-of-antwerp, 2018. Accessed 2 October 2019.

Index

Note: Page numbers followed by *f* indicate figures, *t* indicate tables, and *b* indicate boxes.

the United States
asters